Undergraduate Topics in Computer Science

'Undergraduate Topics in Computer Science' (UTiCS) delivers high-quality instructional content for undergraduates studying in all areas of computing and information science. From core foundational and theoretical material to final-year topics and applications, UTiCS books take a fresh, concise, and modern approach and are ideal for self-study or for a one- or two-semester course. The texts are all authored by established experts in their fields, reviewed by an international advisory board, and contain numerous examples and problems, many of which include fully worked solutions.

The UTiCS concept relies on high-quality, concise books in softback format, and generally a maximum of 275–300 pages. For undergraduate textbooks that are likely to be longer, more expository, Springer continues to offer the highly regarded Texts in Computer Science series, to which we refer potential authors.

More information about this series at http://www.springer.com/series/7592

Kevin Lano · Howard Haughton

Financial Software Engineering

Kevin Lano
Department of Computer Science
King's College London
London, UK

Howard Haughton
Holistic Risk Solutions Ltd.
Croydon, Surrey, UK

ISSN 1863-7310 ISSN 2197-1781 (electronic)
Undergraduate Topics in Computer Science
ISBN 978-3-030-14049-6 ISBN 978-3-030-14050-2 (eBook)
https://doi.org/10.1007/978-3-030-14050-2

Library of Congress Control Number: 2019933710

This Springer imprint is published by the registered company Springer Nature Switzerland AG
The registered company address is: Gewerbestrasse 11, 6330 Cham, Switzerland

Preface

The finance industry continues to diversify and expand and requires the application of increasingly sophisticated software development and data analysis techniques. The use of algorithmic trading and big-data analysis has become commonplace within companies in the sector. Moreover, there is a strong emphasis on the rapid time-to-market of new financial software products and financial models, which can conflict with the achievement of software quality and correctness. There is therefore much interest in considering how these conflicting aspects can be managed and partly resolved, through, for example, the reuse of trusted components, and the use of rapid application development and iterative (agile) development.

In this book, we will introduce the important concepts of the financial software domain and motivate the use of an agile software engineering approach for the development of financial software. We describe the role of software in defining financial models and in computing results from these models. Practical examples from bond pricing, yield curve estimation, share price analysis and valuation of derivative securities are given to illustrate the process of financial software engineering.

The book is intended to support the teaching of software engineering on financial computing courses and can also be used by practitioners interested in adopting agile techniques and software modelling. The material is based on lectures we have given on the Computational Finance programme at King's College London.

London, UK

Kevin Lano
Howard Haughton

Acknowledgements

We would like to acknowledge the contribution of Hessa Alfraihi, Sobhan Yassipour-Tehrani and Shekoufeh Kolahdouz-Rahimi for the creation of the agile model-based development approach and tools used in this book.

About This Book

In Chap. 1, we give an overview of the finance domain and of financial services and markets. In Chap. 2, we consider in more detail some key financial products such as bonds and shares and describe how these are modelled and analysed. Chapter 3 introduces the software lifecycle, model-based development (MBD) and agile development. In Chap. 4, we describe techniques for financial system specification using UML, and in Chap. 5 techniques for financial system design.

Chapter 6 describes the technologies for financial information processing and analysis. Chapter 7 considers the software modernisation and re-engineering. Chapter 8 gives guidelines on how Agile MBD can be practically introduced into development practice. In Chap. 9, we illustrate the process of financial analysis and specification through in-depth study of collateralised debt obligations (CDOs). Finally, in Chap. 10, we describe the tool support for creating financial application specifications and generating code for these specifications on different programming platforms.

In the Appendix, we provide a glossary of terms and exercise solutions.

Throughout the book, we will use a number of case studies based on typical financial engineering problems to illustrate financial software development:

- Internal rate of return calculation for bonds
- Macaulay duration calculation for bonds
- Bootstrapping of interest rates
- Estimation of share price volatility
- Technical analysis of share prices
- Re-engineering Matlab to C#
- Yield curve estimation
- Derivative security pricing
- Risk analysis of CDOs.

The material of these examples can be found at www.nms.kcl.ac.uk/kevin.lano/fse.

Contents

About the Authors

Dr. Kevin Lano worked for over 25 years in the fields of system specification and verification. He was one of the originators of Model-Driven Engineering (MDE) and has been a leading advocate of improving the precision of software modelling and in applying software engineering principles to transformation construction. In recent years, he has worked on the integration of MDE and agile development.

Dr. Howard Haughton has worked in the fields of quantitative finance, risk management and credit risk since 1994. Formally at J. P. Morgan, Dresdner Bank, Deutsche Bank, Merrill Lynch and the Commonwealth Secretariat, he is the director of Holistic Risk Solutions Ltd.

Chapter 1
Financial Services and Markets

In this chapter we give an overview of finance concepts such as financial services, markets and the financial regulatory environment, and discuss the current state of software engineering in finance.

1.1 Introduction

Generally, *financial services* can be described as the means by which financial products are provided to those requiring the product. An example of a financial product would be a bond or a deposit account.

There are a number of key actors involved in the facilitation of a financial service including:

- The service provider. This could be a brokerage house/investment company, commercial bank, insurance company etc.
- The customer/client. This could be an individual, corporate entity, public sector entity etc.
- The regulator. Entities whose objectives are to ensure that service providers treat clients in a manner which does not contravene their rules/guidance and laws, and to ensure that providers maintain certain standards for the benefit of the market as a whole.
- Third party infrastructure providers. This includes those providing trading, clearing and other platforms.

The term *intermediation* is often used in relation to the role played by service providers, in that they act as intermediaries between those with excess funds wishing to invest and those wishing to borrow. As an example, a retail or small investor can invest funds in a unit trust/mutual fund and can benefit from the relatively smaller fees (on a percentage basis) that would be associated with the aggregate amount of such

K. Lano and H. Haughton, *Financial Software Engineering*, Undergraduate Topics in Computer Science, https://doi.org/10.1007/978-3-030-14050-2_1

funds invested. A commercial bank would raise funds by means of offering savings or deposit type products and such funds would then be on-lent to others desiring to acquire financing for a mortgage, for example. An insurance company would invest the premiums received (from those seeking protection against some loss) across a number of investment types and use the returns on these investments to meet their obligations under the insurance contracts.

In the delivery of a financial service it is possible that a number of discrete actions are undertaken by the provider, each or a collection of which constitute what are termed a *transaction*. For example, a new client requesting a brokerage company to trade securities on its behalf might involve:

- Account opening procedures in which clients must provide details of tax/national insurance, passport (or driving license) details, address, income and net worth etc.
- The service provider conducts due diligence on the client in line with the Proceeds of Crime Act, 2002 (POCA) and other relevant regulations
- The client deposits a minimum amount in the brokers account via a funds transfer
- The client provides instructions on the type of securities to be bought/sold, price levels etc.

1.2 Financial Markets

The term *financial markets* refer to the markets in which buyers and sellers trade securities. The term markets also implicitly includes organisations that regulate, infrastructure providers such as the exchanges, as well as the legal framework supporting the market.

There are a number of different types of financial markets including:

- Commodities
- Foreign exchange
- Debt
- Equities
- Derivatives, etc.

The commodities market is generally described as a marketplace for the trading of raw (or primary) products. These products comprise, what are termed, soft or hard commodities. The term soft commodity generally refers to products which are grown and not mined (the latter of which are termed hard commodities). Examples of soft commodities include: cocoa, corn and sugar. Examples of hard commodities include: gold, oil, aluminium.

The foreign exchange market is generally described as the marketplace for the trading of currencies, i.e. the buying/selling and speculation of currencies. It is perhaps worthwhile mentioning that the foreign exchange (or FX) markets are one of the few which are not totally regulated. The spot FX market (i.e. whereby currencies are bought/sold at todays market prices but for delivery in a few days) is unregulated but options and futures contracts on foreign exchange (i.e. where the economics of the contracts depend on future FX rates) are regulated in the United States and there are proposals to introduce European regulations covering similar products.

The debt markets (sometimes known as the credit markets) are generally considered to comprise the marketplace for the trading of bonds. Trading in bonds takes place in one of two markets: primary and secondary. Primary markets are where new bonds are issued for the first time. Such issuances can either be undertaken privately, e.g. where a company raises debt financing by offering a debt security to market participants (either directly or via an underwriter such as an investment company) or issues can be public whereby the debt would be listed on a regulated exchange. Typically the maturity of the debt issued is, normally, greater than one year.

The types of debt issued/traded in the market generally come in two flavours: government or corporate. Government bonds are issued by national/central/federal governments. They are considered risk-free in that the government will *always pay their debt*! In light of the number of sovereign defaults over the last 20 years perhaps the term risk-free is a little misleading. Corporate bonds are issued by entities recognised in law as corporate bodies under the relevant laws of the country in which the entity is incorporated. Since corporate bonds are issued by entities that are perceived to be more risky than sovereigns their bonds, typically, trade at a yield higher than that of a comparable government bond.

The secondary bond market is where previously-issued bonds are resold at some point during their term.

The money markets are, strictly speaking, also a component of the debt markets. The key difference is that the maturity of money market debt is, generally, less than one year. Money markets are considered to be highly liquid, much more than the longer-term debt markets. Common examples of money market products include government treasury bills, commercial paper and repurchase agreements.

The equities markets comprise the marketplace for the trading of equities. Equities are either private or public. Private equity is where a company issues shares in their company directly to investors or via a broker or similar entity. Public equity is where a company has issued shares in the company on a regulated exchange. A fundamental difference between the debt and equity markets is that the latter facilitates for a direct ownership of the company that the shares are invested in.

The derivatives market comprises the marketplace for the trading of derivatives. A *derivative* is a contract between two or more parties whose value is dependent on some underlying asset or index. Examples of derivatives include options, futures and forwards. Prior to the financial crisis of 2008, derivatives were, largely, unregulated and comprised of over the counter (OTC) trading (i.e. direct trading between two parties without an intermediary such as an exchange) and some amount of exchange trades. The lack of regulation was considered to be a contributory factor in the crisis, and the Dodd-Frank Act, 2010 and other similar laws around the world have been introduced to make derivatives much more regulated.

The financial markets have moved on since the days of the ‘barrow boy’ (Fig. 1.1).

Fig. 1.1 "The Old Lady just bought half a yard of cable and there are plenty of bids for Bill and Ben"

The above picture and quote was taken from: http://uk.reuters.com/article/oukoe-uk-markets-slang-forex-idUKTRE80C0LE20120116. The "old lady" refers to the bank of England, a yard is a billion, cable refers to the USD/GBP FX rate and Bill and Ben refers to the Japanese Yen. Thus the quote is saying that the bank of England has bought half a billion pounds against the United States dollar and there is interest for Japanese Yen.

The term barrow boy derives from the working class background of many city traders of yesteryear having worked their way from the back office through to being a star trader. Much trading was conducted via open outcry which is a system in which traders shout out their buy/sell orders. Much of this type of trading has been replaced with that of electronic trading in which traders can enter buy/sell orders electronically and such orders will (hopefully) be filled (either partially or in full) by offsetting orders by others.

1.3 The Legal and Regulatory Context

Laws and regulations exist to ensure that market participants adhere to prescribed rules/laws for the protection of all parties engaged in financial transactions. As with most laws/regulations they act as deterrents but also facilitate for criminal/civil action in the eventuality that they are breached.

1.3.1 The Basel Standards

Significant capital adequacy standards have originated from the Bank of International Settlements (BIS) which have been adopted by many member countries and hence become regulations at the country level.

Basel I: The Basel Capital Accord

The 1988 accord called for a minimum ratio of capital to risk-weighted assets of 8% which was to be implemented by 1992. This accord focussed primarily on credit risk. Credit risk is the risk that a borrower of funds might default on their contractual payments. Another form of credit risk is known as *counterparty risk* where a party to a derivative transaction defaults at a time at which the transaction is in the money to the other party.

By 1991 the '88 accord was amended to allow for the inclusion of provisions in the capital calculations and in '95 and '96 further amendments were incorporated to incorporate bilateral netting and multilateral netting. In '96 amendments were also made to include market risk amendments to the accord to take effect by 1997. The amendment allowed banks to make use of internal models (e.g., Value At Risk (VaR)) to determine their market risk capital requirements. Market risk is the possibility that an investor experiences losses due to changes in market rates (or other factors), such as those of interest, foreign exchange etc. on positions that they hold in securities whose values depend on those rates/factors.

Basel II: The New Capital Framework

The 2004 revised capital framework detailed the three pillars:

1. Minimum capital requirements
2. Supervisory process
3. Disclosure.

Basel 2.5/3

In January 2009 the principles for sound liquidity risk management and supervision were issued. Later in 2009 further documents were issued covering securitisations, off-balance sheet vehicles and trading book exposures. These principles were introduced in light of the inadequacies in regulations observed during the credit crisis of 2008, when many banks were discovered to have de-facto capital/asset ratios substantially below the 8% regulatory minimum, due to the use of mechanisms to remove assets from their published balance sheets.

In 2010 the Basel III standards were issued covering (amongst other aspects): counter-cyclical credit buffers, and leverage ratios. A key idea behind these enhancements was to ensure that banks and internationally active financial institutions retain sufficient common equity Tier 1 capital. In the absence of sufficiency of capital, restraints on capital distribution could be imposed on the bank. Tighter definitions of capital were introduced in 2012/2013 and the updates continue.

1.4 Risk Management and Portfolio Risk

Supply and demand (as well as market manipulation) gives rise to risks within trading portfolios. Changes in macro-economic parameters as well as changes in credit quality also gives rise to risks within banking and trading book portfolios.

Risk management seeks to reduce the effect of losses due to changes in the underlying factors which affect value. In this sense risk can be viewed as downside risk, i.e. where losses occur when factors change. Increasingly financial institutions seek also to exploit opportunities to monetise gains when things move in the "right" direction.

A commercial bank borrows money from retail and corporate clients (i.e., from savers or depositors). They then on-lend portions of these borrowed monies to others for a multiplicity of purposes. For example one type of loan might be for a residential real-estate purchase another might be for a car purchase and yet another might be to facilitate capital investment for a corporate entity.

The risks on the three loans mentioned above are not the same. A lending institution might wish to perform a credit risk assessment (e.g. a credit score or credit rating) on each borrower which reflects the underlying collateral/security for the loan as well as systemic risk (i.e. market factors such as interest and FX rates) and idiosyncratic risk, e.g. default risk of the borrower.

As successive loans are added to the lenders portfolio a picture will emerge as to the composition of the borrowers in the portfolio. Different compositions will result in different risk profiles and hence risk-weighted assets. For a banking portfolio certain hedges might be put in place such as credit default swaps (CDS) (i.e. a product which provides a payment in the eventuality of default of a specified entity), insurance, and other provisions, but such hedges are not likely to be dynamically changing (i.e. selling and buying of the hedges) if the portfolio is largely held to maturity. However, if a portfolio is a trading portfolio then hedges might be bought and sold at a fairly rapid rate depending on the volatility of the market factors influencing value in the portfolio.

A financial institution might want to be able to quantify the risk in a portfolio by analysing the incremental effect of adding a new transaction to the portfolio. Risks can be analysed on the basis of sector/industry, currency, borrower, country etc.

1.5 The Curse or Value of Excel

Many financial models are developed in Excel and these are used to produce valuations of financial products. Excel has also been used by many banks to analyse the risks of investment/banking portfolios. Ideas are prototyped with spreadsheets, and (in many cases) these spreadsheets may form the basis of mission-critical applications.

The benefit of Excel lies in its simplicity, and the vast library of built-in functions with many off-the-shelf packages available to include in the tool. It sits on everyone's desk. Users can even write code in VB, procedurally or using OO style ...what's not to like about it? However, security is compromised as many models built in Excel sit on the PC of the user that created them, and a corruption of the disk could lead to a loss of data and/or the model itself. Even if Excel models are stored on a shared server data integrity is still an issue. Recall that each Excel model is part of autonomous spreadsheets which might "communicate" with each other, but there is no logical data model with enforced data integrity. The spreadsheets are difficult to assess or verify.

As a consequence increased reliance is placed on the creators (or *gatekeepers*) of the models to provide others with an understanding of the "big picture" that the spreadsheets convey. This implies that operational risk is increased for an institution that makes excessive use of Excel for mission critical applications. At the opposite end of the spectrum financial institutions will both procure the acquisition of trading, middle office and settlement systems from third party vendors as well as develop their own in-house systems. Institutions may question why they should develop in-house systems when they can buy off the shelf.

In order to exploit the benefits of proprietary methodologies, knowledge and processes institutions may develop their own software systems rather than use third party tools. It is not uncommon however, to hear that an institution has similar functionality replicated across a myriad of different systems, each of which has either cost the firm huge amounts to build or to procure and maintain.

Some institutions, e.g. JP Morgan, were able to marry concepts of rapid prototyping via spreadsheet-like features and OO development in their *Kapital* system. In this system users were able to create instruments "on the fly" and in a manner which lends itself to verification. This is very useful for creating novel instruments for emerging business areas, and so leads to scalability and enhanced productivity in a fast changing environment. The Kapital system was developed in Smalltalk, which allowed the use of meta-modelling for detailing the financial and risk models of the system and reflection for models (objects) to value themselves, facilitate data integrity, audit etc. The use of Kapital had substantial benefits for JP Morgan in increasing market share and reducing the time to market of new financial products [1].

1.6 Agile Development Processes

Many financial institutions have dated and poor technology infrastructure, inconsistent data models and fairly inflexible management processes which underserve a dynamically changing sector. Frequently updated and far reaching local and international regulations require institutions to be able to rapidly respond to changing requirements. In many instances institutions have to hire many additional contract and permanent staff to accommodate for these changing needs.

Many institutions are still reliant on traditional staged and waterfall models for software development but the industry is transitioning to the use of Agile development. The transition, however, cannot be achieved overnight and requires organisations to "manage a cultural change" from a predominantly serial approach to one that is iterative and more consultative.

The financial domain is often extremely high-pressured, with considerable competitive advantages gained by the organisation which is the first to introduce a new form of financial product, or to find new ways to exploit existing products. Thus financial engineers and software developers face pressures to deliver products as quickly as possible. Agile development can potentially be a means of achieving faster time-to-market by focussing on the delivery of working software that satisfies key requirements as soon as possible (although the software may not be complete and may not cover all requirements). Correctness and accuracy are also very important factors in financial software, and an agile approach needs to be combined with precise mathematical modelling in order to address these aspects.

1.7 Big Data Analysis

Financial institutions, increasingly, understand that the nature of their business and the way in which they implement their business process means that data integrity and data intelligence is an important issue for them. Big data analysis has and will continue to be a significant area of interest for financial institutions seeking to better understand the trends and relationships embedded within financial data and that of exogenous factors. This will lead to enhanced business intelligence on which new marketing, product development and new growth areas can be identified and undertaken.

Big data analysis should not be viewed as a standalone exercise since it should also be used to:

- Drive the data requirements for the integration of new products in trading and other systems
- Assess the implications of regulations/laws on the business model of an institution. This implies that organisations need to adopt enterprise risk analysis for their business processes
- Assess the implications of business decisions on the risks of an institution across the varied silos with a view towards adopting an enterprise risk management (ERM) framework.

Machine learning is also being introduced in this context to recognise patterns in data and build models (e.g., in a neural net) to make predictions based on these models. This has been used particularly for share price prediction, and to identify patterns of trading behaviour which indicate attempts at market manipulation.

Some organisations are moving towards the use of cloud technology as a trade-off between developing in-house systems and outsourcing data integrity and business continuity.

Summary

In this chapter we have given an overview of the essential concepts of the financial applications domain, and we identified some of the issues with the present state of software development in this domain.

Reference

1. Cincom, JP Morgan derives clear benefits from Cincom Smalltalk (2016), www.cincom.com/pdf/CS040819-1.pdf

Chapter 2
Financial Products and Analyses

In this chapter we describe some of the main financial products in detail (bonds, shares, derivative securities) and explain some of the analyses which can be performed to value the products and to define strategies for trading in them.

2.1 Financial Engineering

A wide range of financial products are traded by financial institutions, such as bonds, shares, derivative securities and various forms of structured products such as collateralised debt obligations (CDOs). For such products the task of financial engineering includes: (i) to model a product in terms of its cash flows, risks, value, volatility, etc., (ii) to evaluate significant properties of a specific product, such as the risk of default in a CDO.

In the first case, an institution may be interested in creating a new product or a variation on an existing product, and wishes to explore how this product could operate in practice and benefit the institution. Mathematical models are constructed and different scenarios executed on the models. Key discoveries such as the Black-Scholes equation for pricing derivative securities [1] enabled institutions to effectively model new products (such as options based on shares) and hence to expand their range of products. The contribution of computational modelling in this case is to provide simulation and other numerical and statistical modelling techniques to help explore the behaviour of mathematical models and their impact on pricing and risk management.

In the second case, efficient algorithms are required to compute the predictions of models for specific products. There may be a trade-off between accuracy and timeliness, as for example in high-frequency trading, and heuristics may be used if exact computational procedures are not available.

In this chapter we will look in particular at four different forms of financial product:

- Bonds: these represent a loan of capital funds from an investor to a bond issuer, in return for a regular coupon payment over a fixed term, and return of the loan at the end of the term.

K. Lano and H. Haughton, *Financial Software Engineering*, Undergraduate Topics in Computer Science, https://doi.org/10.1007/978-3-030-14050-2_2

- Shares: these represent a purchase of a part-ownership stake in a publicly-traded company, a share in the equity of the company.
- Derivative securities: a financial product whose value is based upon the value of an underlying asset, which could be a physical commodity such as gold, or another financial product, such as shares.
- Collateralised Debt Obligations (CDOs): aggregated portfolios of debts.

2.2 Bonds

Bonds are a means for a company or government to raise funds from investors whereby the repayments are based, inter alia, on either a fixed or a floating interest rate, a *principal* figure denoting the amount borrowed, and a *term* over which the funds are paid back. For businesses, bonds have the advantage that investors do not acquire a stake in the company (as in the case of shares). However, the value of the bond as an investment depends upon interest rates: a general increase in interest rates will reduce the value of fixed interest rate bonds, since higher returns will be available elsewhere.

A bond is purchased for a given price including an initial investment amount, known as the *principal* of the bond. Examples of UK government bonds can be found at dmo.gov.uk. When calculating bond values, it is usual to consider a nominal principal amount for which 100 is a convenient figure. A bond has a *term*, eg., 5 years, 10 years, etc. The most common form of bond is a *bullet bond* in which coupon payments are paid to the investor throughout the life of the bond, with the last payment being the sum of the final coupon plus return of the investment amount (redemption of the loan). In a fixed-coupon bond the coupon rate c is defined as an annual percentage of the investment amount, and the coupon is paid at regular intervals (twice per year, or once per year, etc.) during the term.

In terms of cash flows, the initial payment of *price* is a flow from the investor (lender) to the bond issuer (who is the borrower of the investment). The coupon payments are regular cash flows of $coupon/f$ from the borrower to the investor, where *coupon* is the annual coupon amount, expressed as the coupon rate c times the principal (100), and f is the frequency of coupon payments per year. There are $f * term$ payments of the $coupon/f$ amount in total. At term, there is also a cash flow of 100 from the borrower to the investor.

For example, consider an investment of £100 in a 20-year 2% annual coupon bond, with a price of £105. The investor receives coupon amounts £2 at the end of each of the first 19 years, then £102 (coupon plus principal) at the term (Fig. 2.1).

Since the price is greater than 100, the bond is trading at a premium, ie., 105% of the face/principal amount. If the price were less than 100, then the bond would be trading at a discount.

Mathematical modelling of bonds is used to compute their value under different market situations, in particular, how their value depends upon the underlying risk-free interest rate r. The basic quantity for a bond is its *payout*: the actual total amount

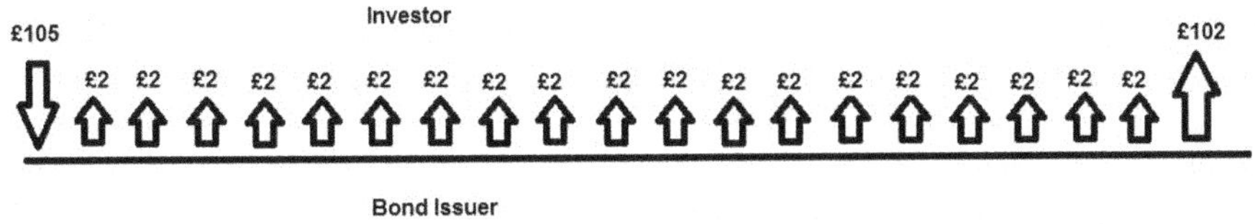

Fig. 2.1 Cash flows for a bond

which an investor receives back. This is the sum of all the coupon payments over the term, plus the initial investment. For fixed coupon bonds with a whole-integer term we have

$$payout = term * coupon + 100$$

or equivalently

$$payout = term * c * 100 + 100$$

This is £140 in the example.

But the value of the money received at a future time (eg., N years in the future) is typically reduced compared to the same amount received today. An amount X received N years in the future, with respect to an annual interest rate r, is worth $X/(1+r)^N$ today. We say that the amount X is *discounted* to $X/(1+r)^N$. $r \geq 0$ is usually assumed, so $X/(1+r)^N \leq X$.

The reason for this decrease in value is that an investor could instead invest their funds at interest rate r, whereby the amount Y today would have grown to $Y * (1+r)^N$ after N years. Thus the amount X in N years time is equivalent to the amount $X/(1+r)^N$ today, given an annual interest rate r.

For a fixed-coupon bond with annual coupon payments, the present value, $value(r)$, of a bond is then the sum of all the discounted payments over the bond term, assuming an effective annual interest rate of r over the term of the bond:

$$value(r) = (\Sigma_{t=1}^{term} coupon/(1+r)^t) + 100/(1+r)^{term}$$

Notice that as r increases, $value(r)$ decreases, ie., the partial derivative of $value(r)$ wrt r is negative. For example, the $value(0.01)$ at 1% interest rate of our 20-year bond is £118. However, $value(0.02)$ is £100.

Since interest rates will vary over the term, it is simpler to work with an effective equivalent uniform rate when calculating value, however it is also possible to compute the value using non-uniform rates, with different r values used for each discounted payment.

The price of the bond already contains an assumption by the bond seller about what the effective interest rate r will be over the term: the price is set so that it is close to $value(r)$. At least, this is the case for government bonds and bonds offered by institutions with a very low risk of defaulting on the loan (failing to repay it). The assumed rate in this case will be close to the risk-free interest rate because the bonds themselves are considered almost risk-free as investments (the UK government has never, yet, defaulted on its bond loans).

Therefore it is interesting to discover the rate r at which the market price of the bond equals $value(r)$. This is called the *internal rate of return (IRR)* of the bond, also referred to as its *yield*, and measures the quality of the investment. The IRR satisfies

$$value(irr) = price$$

The investment is profitable for the investor if the IRR is higher than any other available yields of investments of the same term and of equivalent or lower risk, i.e., the bond returns more value than an alternative investment of comparable risk. A high IRR indicates that the bond is more likely to be profitable for the investor (in the absence of defaults): the bond is profitable wrt an alternative investment with yield r whenever $irr > r$, therefore a larger value of irr means that the bond remains profitable wrt a wider range of available alternative investments, compared to a bond with a lower irr.

Looked at in a different way, a bond with a high IRR may be under-priced and hence a good investment. The IRR of our example is 0.017, ie., 1.7%. In general, computation of the IRR requires the use of some iterative approximation procedure such as the secant or bisection techniques. We look at ways to solve this problem in Chap. 4.

Interest rates are not the only factor in bond prices. The risk of default of the bond seller (borrower) also has to be considered: in general, the more risky the bond is, the higher the coupon rate offered by the borrower, and hence their bonds will have a higher IRR and/or lower price compared to a less risky bond (such as government gilts). The risk of default on a loan is classified by debt ratings agencies in categories ranging from AAA (the lowest risk, essentially risk-free), to AA, A, BBB, BB, B, CCC. BBB grade to AAA are called ‘investment grade’, whilst below BBB is called ‘junk grade’. The rating of a company or sovereign state has high impact on their ability to raise money through bonds: the loss of AAA rating for a state will significantly increase their costs of raising money if the rating falls substantially. Finally, bonds are also subject to market forces, which means their price will normally rise if there is a high demand for them, since interest rates will fall.

A *zero-coupon* bond pays no coupons during its term, but only an accumulated interest amount together with the principal repayment at the end of the term. The valuation of these bonds is quite simple, because the $value(r)$ equation becomes:

$$value(r) = R/(1 + r)^{term}$$

for discrete compounding, where R is the total repayment on 100 nominal principal. The IRR can then be directly calculated as

$$irr = (R/price)^{\frac{1}{term}} - 1$$

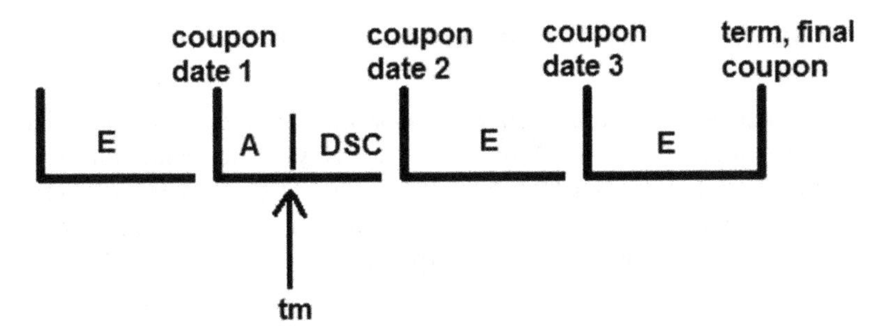

Fig. 2.2 General bond trading situation

Some bonds may be traded during their term. The value of a bond at a point *tm* during its lifetime can be calculated based on the remaining future cash flows due from it (where *tm* is an integer and falls on a coupon payment date):

$$value(r, tm) = (\Sigma_{t=1}^{term-tm} coupon/(1+r)^t) + 100/(1+r)^{term-tm}$$

in the case of an annual fixed coupon bond, with *tm* an integer, $tm \geq 1$, $tm < term$.

The situation is more complex if *tm* falls between coupon dates, in that case we need to take account of the *accrued interest* for the next coupon, and the displacement of the purchase date *tm* from the payment dates. Let *DSC* be the number of days from settlement (ie., the date *tm* when the bond is bought/sold) until the next coupon date, and *E* the normal number of days in the coupon period. Eg., for a twice-yearly coupon, *E* could be 182, half of the number of days per year. Let *A* be $DSC - E$, the number of days from the beginning of the coupon period to the settlement date (Fig. 2.2).

Then the accrued interest for the next coupon payment is

$$accruedInterest = \frac{coupon}{frequency} * \frac{A}{E}$$

taking *coupon* as the principal times the coupon percentage, and *frequency* as the number of payments per year.

The valuation is modified to:

$$value(r, tm) = \left(\Sigma_{t=1}^{N} \frac{coupon}{frequency} \Big/ \left(1 + \frac{r}{frequency}\right)^{t-1+\frac{DSC}{E}} \right) + \\ Face \Big/ \left(1 + \frac{r}{frequency}\right)^{N-1+\frac{DSC}{E}} + accruedInterest$$

where *N* is the number of remaining future coupon payments after *tm*, and *Face* is the face amount of the bond, usually 100.

On a coupon date, $A = 0$ and $DSC = E$, so that this version of $value(r, tm)$ reduces to the previous case for annual coupon bonds.

2.2.1 Discrete Versus Continuous Compounding

The above calculations used *discrete compounding* of interest rates: an amount X grows to an amount $X * (1 + r)^N$ if invested for N years at annual rate r, where the increment $r * amount$ is added to $amount$ once at the end of each year. An alternative is to continually apply r (in practice this amounts to compounding at a daily frequency). Mathematically, this is expressed as

$$X * e^{r*N}$$

If we use continuous compounding then the discounting computation becomes $X * e^{-r*N}$ instead of $X/(1 + r)^N$.

2.2.2 Yield Curves

The professional valuation of bonds makes use of the predicted interest rates over the term of the bond. The variation of interest rates with the length of an investment is known as the *term structure of interest rates* or a *yield curve* (Figs. 2.3 and 2.4 show example yield curves). The x-axis represents the term of the investments in years, the y-axis shows the yield of the investments: assuming that these commence from the current date.

The basis of estimating a yield curve is to identify a set of market data points (ie., bond prices or yields) from a range of financial assets, with similar risk levels, within one country. Typically these are based on government bonds of varying terms for the specific country. By computing the effective duration and IRR of bonds of different durations, a set of $(time, rate)$ points are obtained, and a computational optimisation procedure can be applied to fit a curve to these data points, and hence enable a rate to be assigned to any duration. Different models exist (eg., the Nelson-Siegel formula [2]) for the shape of the curve. The Nelson-Siegal (NS) model for yield curves uses the formula

$$y(t) = \beta_1 + \beta_2 * (1 - exp(-t/\lambda_1))/(t/\lambda_1) +$$
$$\beta_3 * ((1 - exp(-t/\lambda_1))/(t/\lambda_1) - exp(-t/\lambda_1))$$

to compute the yield $y(t)$ for given duration t, where β_1, β_2, β_3 and λ_1 are real-valued constants. Figures 2.3 and 2.4 are yield curves that follow this equation.

The curve has a long-term rate component (β_1), short-term component (2nd factor), and a 'hump' (3rd factor). $\beta_1 > 0$ is assumed, as is $\beta_1 + \beta_2 > 0$ and $\lambda_1 > 0$. The problem is to estimate the parameters β_i and λ_1 which make the curve best fit a set of given market data points: this process is termed 'fitting the curve' to the data. The Nelson-Siegal-Svensson (NSS) model adds a further 'hump' term with parameters β_4 and λ_2.

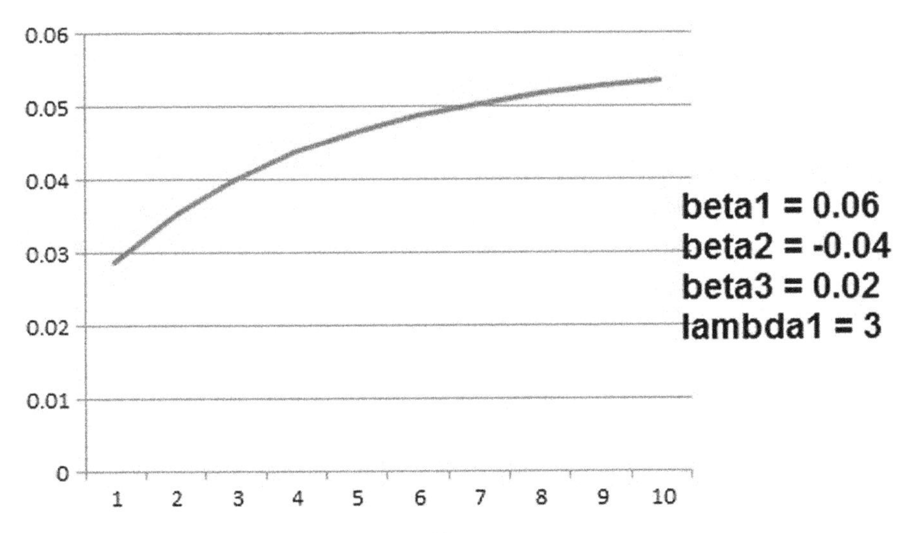

Fig. 2.3 Example Nelson-Siegal yield curve (1)

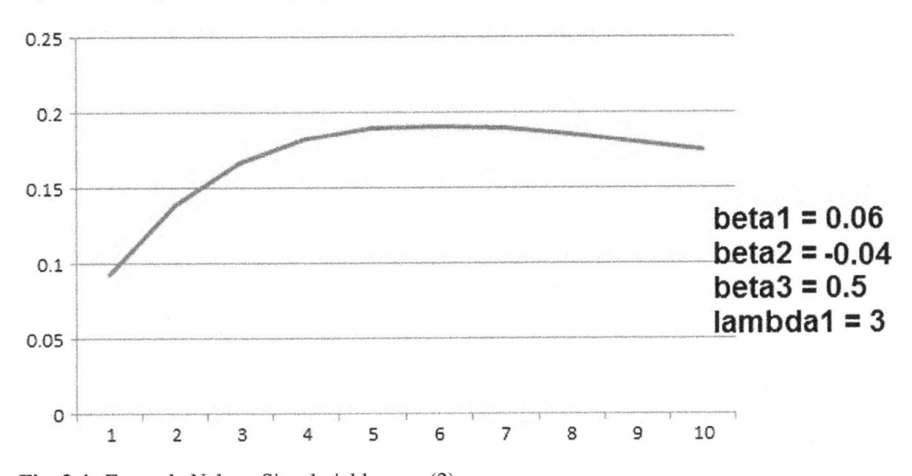

Fig. 2.4 Example Nelson-Siegal yield curve (2)

In Chap. 7 we will implement yield curve estimation using numerical optimisation techniques.

From a yield curve the price of a bond of a particular duration d can be calculated by finding $y(d)$ and using this to compute $value(y(d))$ for the bond.

Yield curves tend to be upwards sloping because investors prefer more liquid (shorter term) investments where possible, and hence require higher yields on longer-term investments. In addition, longer-term investments carry a higher risk of default and hence again a higher yield is required in order to compensate for this risk. Inverted yield curves (where short-term yields are higher than long-term) can indicate a lack of confidence in the economic future, and strongly inverted curves have historically preceded recessions.

2.2.3 Floating Rate and Index-Linked Bonds

A floating-rate bond has coupons which vary from payment to payment based on an underlying interest rate, such as the LIBOR (London Interbank Offered Rate: the rate that banks lend to each other). This type of bond offers some protection against interest rate rises which would decrease the value of a fixed-rate bond. However, the varying rate of coupons is a potential disadvantage. A floating-rate bond contract specifies the *reset period*, ie., how often the payment rate is adjusted. It is also possible to obtain bonds which start at a fixed rate for a given period and then transfer to a floating rate.

Typically, the rate $r1$ to be applied at the next coupon date will be known, but not subsequent rates, if coupon and reset dates coincide. Eg., for semi-annual resets and coupons following LIBOR + 10 basis points, $r1$ would be the 6-month LIBOR rate at the preceding reset point, plus 0.1%. The next coupon is the principal (eg., 100) times the current floating rate $r1$: $coupon = 100 * r1$. The value at a time interval t before the next coupon date is then:

$$value(r) = 100 * (1 + r1) * e^{-r*t}$$

where r is the relevant continuous compounding rate applicable to period t.

Index-linked bonds also have varying cash flows. Such bonds link the coupon rates to some price index/inflation index in order to cancel out the effects of inflation on the value of coupon payments. For example, a UK index-linked gilt has coupon payments and principal payments adjusted in line with the retail price index (RPI), equivalent to the consumer price index in the US.

A simplified version of this process is to consider that at a date t, the indexation factor $ifact_t$ is computed as the ratio of the index rate at time t compared to the rate at the issue date s of the bond: $ifact_t = index_t/index_s$. The principal is then adjusted to $100 * ifact_t$ and coupons are recalculated based on this adjusted principal:

$$coupon_t = c * 100 * ifact_t$$

Thus the valuation formula at time point t becomes:

$$value(r) = \left(\Sigma_{i=1}^{term} coupon_t/(1+r)^i\right) + 100 * ifact_t/(1+r)^{term}$$

for an annual repayment index-linked bond with $term \geq 1$ years remaining until the bond matures. This value is $ifact_t * value_0(r)$ where $value_0(r)$ is the usual valuation formula

$$value_0(r) = \left(\Sigma_{i=1}^{term} c * 100/(1+r)^i\right) + 100/(1+r)^{term}$$

Note that *value(r)* is an approximation, since the actual indexation factor that will be applied to future coupon payments is not known at time t.

2.3 Shares and Stocks

Shares represent a part share ownership in the equity of a company: the surplus of its assets over its liabilities. Selling shares in its equity is another means for a business to raise funds. Shares tend to be more volatile than bonds, and hence can be viewed as riskier as an investment. Apart from trading gains and losses, shares can also provide *cash dividends* paid as a share of a company's profits to its shareholders, with the amount paid proportional to their share holding. *Share dividends* involve a company issuing more shares to its existing shareholders, eg., one new share for every 4 already held. This can be analysed as subdividing the existing shares.

Share price analysis is concerned with the possible loss or gain in a portfolio of shares over different periods of time, and with the choice of trading strategies in order to maximise gain. *Technical analysis* seeks to use past performance data to predict future performance, whilst *Fundamental analysis* seeks to evaluate key properties of a business as a guide to its share performance. Both kinds of analysis use indicators for trading, to determine appropriate sell/buy/hold actions for particular shares. There is a sense in which technical analysis is futile, because if it were possible to predict future share prices on their past history, then the same predictions would be read by many investors, and their consequent concerted selling/buying behaviour would remove any advantage from the prediction.

Indeed, a fundamental assumption in the usual mathematical analysis of shares is that their prices follow a Markov process, that is, the future price depends only on the current price and not on any previous history. Notwithstanding these limitations, different heuristic or algorithmic approaches have been used for technical analysis share price prediction: neural nets; genetic algorithms; K nearest-neighbours, etc.

Historical share information is also used to estimate the volatility of a share (how much its price is likely to change) and the correlation of different shares (or market sectors). Diversification of share portfolios to include combinations of shares with negative correlations can significantly reduce overall volatility and reduce the risk of losses.

2.3.1 *Share Trading Processes*

Most stock markets manage trades in shares via a system termed *continuous double auction*, whereby orders to sell and buy shares are submitted to the exchange and either executed immediately, or placed in a queue if the order specifies a limiting price for sale/purchase which cannot yet be achieved in the current state of the market.

The first type of order is termed a *market order*, and identifies the company to be traded, the volume of shares to trade, and the position (buy or sell). Limit orders also specify a limit price: for purchase this is a maximum limit (the *bid* price) on the purchase price, while for sell orders it is a minimum limit price (the *ask* price) for a sale.

Limit orders which cannot be executed directly will be queued in a *limit order book*, which ranks the orders by the limit prices: the bid orders ranked from the highest to the lowest limit prices (ie., the orders most likely to be satisfied are ranked above ones less likely to be satisfied), and the ask orders ranked from the lowest to the highest limit prices. When a matching order is received for a queued order, a trade is completed, ie., a transaction of sale and purchase of the same quantity of the given share.

The state of the limit order book is usually publicly visible to all traders, and this can give rise to risks of market manipulation, since orders may be cancelled or amended while in the queue. Attempts at market manipulation can consist of submitting orders which are intended to affect the share price, and are cancelled before they can be executed.

2.4 Derivative Securities

Derivative securities include a wide variety of financial instruments whose value depends upon an underlying asset. For example, *forward contracts* are an agreement to buy or sell an asset at a fixed price at a fixed future time. These are generally traded over the counter as a private agreement between two parties. In contrast *futures contracts* permit variation of the maturity date within some limits, and are traded on an exchange and tend to be standardised. *Options* give the holder of the option the right (but not the obligation) to buy (call option) or to sell (put option) an asset on or before a date for a fixed price. *American* options can be exercised at any time up to the expiration date, whilst *European* options can only be exercised on the expiration date.

Derivative securities can be based on almost any underlying variable: (i) traded securities such as shares, bonds, gold; (ii) financial variables such as interest rates, exchange rates or prices of general commodities; (iii) other variables such as climatic variables, election results, etc.

Derivative securities can be used for *hedging*: limiting the losses from a contract or financial position, or for *speculation*: using the derivative to reduce the amount of funds needed to speculate on an asset, ie., to increase the leverage of the investor's funds.

Options have become an essential feature of financial markets since stock options were first traded on an exchange in 1973. Consequently the valuation of options and other derivative securities is of major importance. The valuation of derivative securities takes into account the value of the underlying asset, the underlying interest rates, and other factors. In some cases a precise formula for the value has not yet been found. We consider techniques for derivative security valuation in Chap. 8.

2.5 Collateralised Debt Obligations

A collateralised debt obligation (CDO) is an aggregated set of debts (such as mortgage loans or business loans made by a bank), the cash flows from which are then used as a basis for further financial products. This is termed *securitisation* of the loans. The aggregation of debts with different (and presumably uncorrelated) default characteristics has the benefit (in principle) of reducing the overall impact of defaults. The analysis of the risk associated with CDOs is highly critical for the institutions that trade in them: the global financial crisis of 2008–2009 was in large part caused by incorrect estimations of risks for CDOs based on sub-prime mortgage loans.

Figure 2.5 shows an example of a CDO with underlying collateral loans and bonds owned by a bank X. Loans are created whenever X lends to borrowers, e.g. for a business loan. From the perspective of the bank, the loan is an asset which it is using to earn income (a cash flow of interest payments). However, each borrower has a probability of default, related to its credit rating. So each loan/bond of X has a default probability. The overall risk of loss from individual and multiple defaults can be computed from individual default rates and assumptions about the correlation of defaults.

To set up a CDO from a collection of debts, the bank would establish a special purpose vehicle (SPV), probably incorporated in a tax advantage jurisdiction like the Cayman Islands, which is considered bankruptcy remote to bank X. This ensures that investors on the notes/securities issued by the SPV have no recourse to X if there are defaults with the securities issued by the SPV. The SPV issues securities backed from the cash flows from the underlying loan pool (eg., mortgage payments, loan interest payments). A typical loan pool might consist of several hundred up to thousands of loans used for backing a single security issued by the SPV. The issued securities are collateralised by the cash flows from the more primitive loan pool.

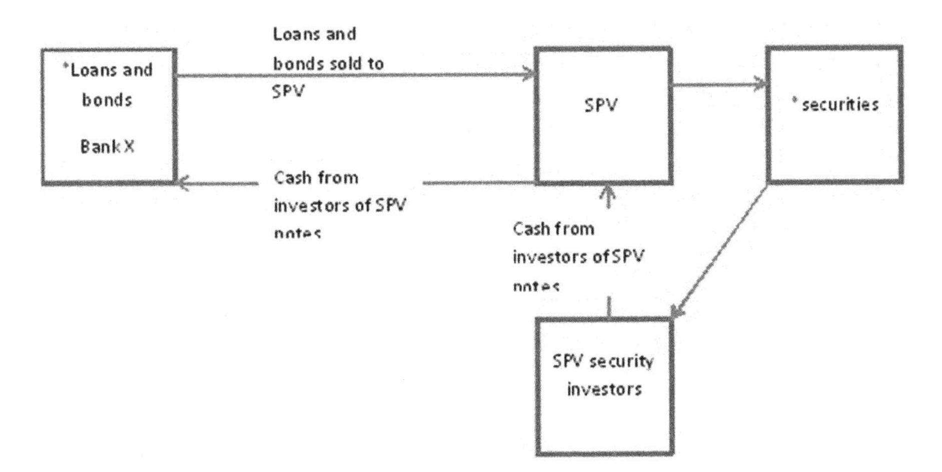

Fig. 2.5 CDO example

The cash obtained from the investors are used to facilitate bank X undertaking more loans and bonds and changing the structure of its balance sheet and risk-weighted assets and capital associated with those assets. In particular, the economic risks of the assets are removed from the balance sheet of X, meaning that it gains the ability to re-lend the asset amount without violating the Basel regulations requiring a certain percentage of its assets to be retained as equity (eg., 8%). This increases the effective leverage ratio of the bank (the ratio of its assets to its equity).

The analysis of CDOs involves computing the risk of losses from single and multiple defaults within the aggregated pools of loans, some of which may be (partly) correlated and others assumed to be uncorrelated. If there are n loans in a pool of loans owned by bank X, we need to find the probability that m from n number of defaults will occur where $m \leq n$. This information is useful for a bank in understanding the risks it faces from borrowers defaulting. The information may also be used by rating agencies to determine the credit rating of securities issued by the SPV.

One means to organise the CDO is to group loans into sectors, eg., all loans to other financial institutions in one sector, loans to oil and gas businesses in another sector, mortgage loans in another, etc. The assumption is made that defaults within one sector cannot cause defaults in another, but only possibly cause defaults within the same sector (termed 'default infection').

Risk contributions allow a bank to be able to assess the relative riskiness of one sector with respect to another. If there is a downturn in the economy which affects one sector more than another then a bank might be more inclined to lend to companies in the least affected sector more than in the most affected sectors. In Chap. 9 we carry out analysis of CDO risks and evaluate the risk contributions of particular borrowers and sectors within a CDO.

2.6 Essential Calculus

In subsequent chapters we will make use of some basic facts of calculus. Properties of differentials arise in financial theory since many finance processes can be understood based on limits $\frac{dV}{dt}$ of discrete changes ΔV in variables over discrete time intervals Δt. In addition, numerical estimation and optimisation procedures such as the secant or Newton-Raphson techniques are based on calculus concepts.

Recall that if f, g are differentiable, and so are the composed functions $f + g$, $f * g$ and f/g, then the composed functions have differentials:

$$\frac{d(f+g)}{dt} = \frac{df}{dt} + \frac{dg}{dt}$$

$$\frac{d(f*g)}{dt} = g * \frac{df}{dt} + f * \frac{dg}{dt}$$

$$\frac{d(f/g)}{dt} = (g * \frac{df}{dt} - f * \frac{dg}{dt})/g^2.$$

The exponential function has the key property that it is its own differential:

$$\frac{de^t}{dt} = e^t$$

From these results we can infer, for example, that the change with time of the present-day value $X * e^{-r*t}$ of an amount X received time t in the future subject to rate r

$$\frac{d(X * e^{-r*t})}{dt}$$

is always negative, for $X > 0$ and $r > 0$, ie., $X * e^{-r*t}$ decreases with increasing t.

Partial differentiation arises when a function of several variables is differentiated with respect to one of these. For example, $X * e^{-r*t}$ may be considered as a function of both r and t. Its partial differential wrt r is written as

$$\frac{\partial X * e^{-r*t}}{\partial r}$$

This expresses how the value function varies wrt r, and shows that this also decreases as r increases.

A particularly important partial differential equation is the Black-Scholes equation:

$$\frac{\partial f}{\partial t} + r * S * \frac{\partial f}{\partial S} + \frac{1}{2} * \sigma^2 * S^2 * \frac{\partial^2 f}{\partial S^2} = r * f$$

This describes how the value f of a derivative security based on an asset with value S varies wrt time t and wrt S. r is the risk-free interest rate, and σ the volatility of the underlying asset. Different solutions to this equation provide valuation formulae for many kinds of derivative security, as we discuss in Chap. 8.

2.7 Combinatorics and Statistical Properties

Combinatorics concerns quantifying the number of different ways that an event can occur or a choice can be made. The number C^n_r for natural numbers n and r with $n \geq r$ is referred to as "n choose r" or as the binomial coefficient of n over r. It represents the number of different ways that a set of r elements can be chosen from a set of size n. For example, 2 elements can be chosen from a set $\{a, b, c, d, e\}$ in essentially $\frac{5*4}{2} = 10$ different ways (5 choices of the first element, then 4 for the second, then discarding the duplicated choice sets).

Mathematically, C_r^n is

$$n!/(r! * (n - r)!)$$

However, it is more efficient to compute C_r^n as

$$(\Pi_{i=n-r+1}^n i)/r!$$

Eg., $C_2^5 = 4 * 5/2 * 1$. Whichever is the smaller of $n - r$ and r is used as the lower argument of C_r^n, since $C_r^n = C_{n-r}^n$.

Statistical distributions also play an important role in financial analysis, together with properties of distributions such as their average or *mean* and their *standard deviation*.

If a series of n measurements of some quantity (eg., a share closing price on successive trading days) is made, the result can be expressed as a discrete distribution $d = Sequence\{x_1, ..., x_n\}$ where the x_i are ordered from low to high values.

The average $mean(d)$ of d is simply $(\Sigma_i x_i)/n$ or $d \rightarrow sum()/n$ in OCL notation (Chap. 4). The notation μ_d can be used for the mean.

The *variance* of d measures how much the x_i are dispersed from their mean point. Variance is denoted σ_d^2 and defined as:

$$\sigma_d^2 = (\Sigma_i (x_i - \mu_d)^2)/n$$

The *standard deviation* is the positive square root σ_d of the variance.

In the example of share price data, σ_d gives an indication of how volatile the prices are over a time period. The symbol σ is usually used to denote volatility in the financial context.

The mean and variance are termed the first and second *moments* of a distribution. Higher moments (based on higher powers of $x - \mu$ for a random variable x and its mean μ) include *skewness*, a measure of how unsymmetrical a distribution is, and *kurtosis*, a measure of dispersion of the distribution.

A collection of measurements such as d can be regarded as a discrete statistical distribution with probability density function *pdf* defined by:

$$pdf(x) = d \rightarrow count(x)/n$$

That is, the probability of x is the number of times it occurs in d, divided by the size of d.

A property of any *pdf* is that

$$\Sigma_x pdf(x) = 1$$

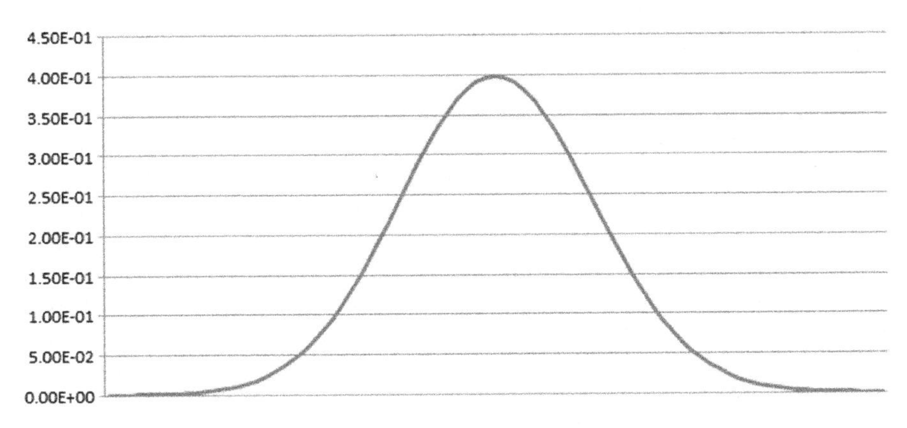

Fig. 2.6 Normal distribution $N(0, 1)$

Related to the *pdf* is the *cumulative density function, cdf*. The value of $cdf(x)$ is the proportion of the distribution which is $\leq x$. For a discrete distribution d:

$$cdf(x) = \Sigma_{y:d\&y\leq x} pdf(y)$$

An important concept for any distribution or random variable is the *expectation* or *expected value* of the distribution/variable. This is the probability-weighted sum of the possible outcomes:

$$E(V) = \Sigma_x prob(V = x) * x$$

For the discrete distribution d, $E(d) = \Sigma_i pdf(x_i) * x_i$, which is the mean μ_d. In finance the expectation is also often denoted by μ (eg., the expected return from an investment), but can be applied to any outcome where probabilities can be assessed. For example, if you enter into a contract which has a 99% chance of earning £1000 profit, and a 1% chance of losing £1,000,000, your expected outcome is a loss of £9010.

A key distribution both in finance theory and other domains is the *normal distribution* (Fig. 2.6). This has the *pdf* defined by:

$$pdf(x) = e^{-0.5*x^2/\sigma^2}/(\sigma * \surd(2 * \pi))$$

where σ is the standard deviation. The mean is 0. This standard distribution is denoted $N(0, \sigma)$. The pdf has the property that $pdf(x) = pdf(-x)$ because the distribution is symmetric about 0.

The *cdf* cannot be defined by a closed formula, but approximations can be used instead. The *cdf* of the normal distribution arises in the solutions to the Black-Scholes equation in particular (Chap. 8).

With reference to a normal distribution $N(0, \sigma)$, the overall probability of a randomly sampled point x from the distribution being in the range $(-\sigma, \sigma)$ is 68.3%, of being in the range $(-2 * \sigma, 2 * \sigma)$ is 95.4%, and of being in $(-3 * \sigma, 3 * \sigma)$ is 99.7%. The term "3-sigma event" is used to refer to an event which is as unlikely as obtaining a $N(0, \sigma)$ sample outside the range $(-3 * \sigma, 3 * \sigma)$. That means the event should occur no more than 3 times out of 1000. Stronger restrictions are "4-sigma event","5-sigma event", etc. Such a designation (and probability) only makes sense if the event in question *actually* follows the normal distribution. The distribution is only a mathematical approximation for the real behaviour of such variables as stock-exchange indexes. Recent history of the financial world has shown that this assumption can be misleading for cases of exceptional events such as stock market crashes and other financial crises!

The normal distribution is also used in the mathematical modelling of share prices as a stochastic process. For example the share price S can be modelled as a solution of an equation

$$\frac{dS}{S} = \mu * dt + \sigma * \eta * \sqrt{dt}$$

where μ is the expected rate of return of the share, and σ the volatility (it is the standard deviation of the proportional change in S over time), and η is a sample from $N(0, 1)$. This model or other similar stochastic models can be used for *Monte-Carlo simulation* of share prices by generating large numbers of samples η to represent possible price movements of the share.

Another distribution which arises when considering the random arrival of rare events (such as defaults on loans) is the *Poisson distribution*. This has *pdf* defined as:

$$pdf(x) = e^{-\mu} * \mu^x / x!$$

for positive integers x, where the mean μ is also the variance.

Correlation and *covariance* are also important statistical properties of financial quantities, providing a measure of how closely linked two or more separate quantities are. If V_1 and V_2 are two random variables (eg., representing defaults in two separate loans, with $V_i = 1$ indicating default of i and $V_i = 0$ indicating no default), then it is interesting to determine the expectation $E(V_1 * V_2)$ of both loans defaulting, given known expectations $E(V_1)$ and $E(V_2)$ of individual defaults. In general, $E(V_1 * V_2) \neq E(V_1) * E(V_2)$, although the equality holds if the variables are statistically independent.

The quantity

$$C(V_1, V_2) = E(V_1 * V_2) - E(V_1) * E(V_2)$$

is termed the *covariance* of V_1 and V_2. A value of 0 indicates that the variables are uncorrelated (although they may not be statistically independent).

The values for covariance can be derived from historical data or from theory. A negative covariance indicates that the variables tend to change value in opposite

Table 2.1 Covariance matrix

	V_1	V_2
V_1	$p * (1 - p)$	$p * (p - 1)$
V_2	$p * (p - 1)$	$p * (1 - p)$

directions, a positive covariance indicates that they tend to change values in the same direction.

For example, if $V_2 = 1 - V_1$ then $C(V_1, V_2) = p * (p - 1)$ where p is the probability of $V_1 = 1$. On the other hand, if $V_1 = V_2$, then $C(V_1, V_2) = p * (1 - p)$.

The covariances of different variables can be represented in a covariance matrix, which has an entry for each pair of variables. This is symmetric about the axis, which contains the variances of individual variables. In the case of $V_2 = 1 - V_1$ the matrix is as shown in Table 2.1.

A related concept is the *correlation coefficient* between two random variables, which expresses the strength of their correlation. This is defined as

$$\rho(V_1, V_2) = C(V_1, V_2)/\sqrt{}(\sigma_1^2 * \sigma_2^2)$$

where σ_i^2 is the variance of V_i. For the first example above, $\rho(V_1, V_2) = -1$, indicating a complete inverse correlation between the variables. For the second example the correlation is 1.

In general the correlation coefficient ranges between -1 and 1. When it is positive this indicates that the variables tend to increase or decrease in value together. When negative, this indicates that as one increases the other tends to decrease, and vice-versa.

In portfolio management the covariance between the value of two or more financial products is a key element in reducing overall variance (volatility) and increasing the overall value of the portfolio. One reason for this is that:

$$variance(V_1 + V_2) = variance(V_1) + variance(V_2) + 2 * C(V_1, V_2)$$

so that a negative covariance will reduce the variance of a sum of two variables compared to the sum of their variances. In our first example, the variance of $V_1 + V_2$ is reduced to 0 because the sum is constantly 1.

Summary

We have given an overview of some important financial products and their analysis. We have also explained essential calculus and statistical theory which underpin financial analysis.

We will use examples of financial analysis in subsequent chapters to show how different software engineering techniques and underlying technologies can be applied to financial software problems.

Exercises

1. In Sect. 2.2, what does *value*(0) represent?

2. Given two bonds, both with term 10 years from the same date, annual coupon rate 2% and annual payments, but with different prices £105 and £110, determine without calculation which has the higher yield.

3. How does the $value(r, tm)$ of a bond change as *tm* approaches the *term* of the bond?

4. Generalise the definition of $value(r)$ to $value(rs)$ where *rs* is a sequence of interest rates of length *term*, $rs[i]$ represents the annual interest rate in year *i* of the investment, $i = 1$ to *term*.

5. A *parallel shift* of a yield curve represents a change in the curve where the rates for all time periods move by the same amount δ up or down. What parameter(s) of the NS equation can be used to express such a change?

6. Risk assessment of CDOs uses segregation of the underlying loans into different sectors, with the assumption that defaults in one industrial/business sector are independent of defaults in another. Is this assumption always valid? Suggest two possible counter-examples where a company failure in one sector can cause failures in another.

7. Compute the value of a resold bond according to the model of Fig. 2.2 where the frequency is 2, annual coupon is 5.75%, DSC is 90, E is 182, A is 92, the term is 10 years and N = 20.

References

1. F. Black, M. Scholes, The pricing of options and corporate liabilities. J. Political Econom. **81**, 637–659 (1973)
2. C. Nelson, A. Siegal, Parsimonious modeling of yield curves. J Business **60**(4), 473–489 (1987)

Chapter 3
Model-Based and Agile Development

This chapter introduces the main concepts of the software development lifecycle, and describes software specification techniques and development approaches which can be used for financial applications:

- Software modelling using UML
- Model-based development (MBD)
- Domain-specific modelling
- Agile development methods: Scrum, Kanban and XP.

The chapter will give a non-technical overview of these topics, and they will be developed in more detail in subsequent chapters.

3.1 The Software Development Lifecycle

The following activities take place in any software development process, whether organised into strict sequential stages (as with the classical 'waterfall' process) or iterated in repeated cycles (as in agile methods such as Scrum). The software lifecycle stages are:

- Feasibility analysis
- Requirements analysis
- Specification
- Design
- Implementation
- Testing
- Maintenance and Decommissioning.

K. Lano and H. Haughton, *Financial Software Engineering*, Undergraduate Topics in Computer Science, https://doi.org/10.1007/978-3-030-14050-2_3

3.1.1 Feasibility Analysis

This stage asks whether there is a business case for the system and if it will actually be used. It considers (i) Technical feasibility—is it possible to build the system with the available technology? (ii) Financial feasibility—can it be developed with the available budget? (iii) Time—is it possible to develop in a useful time-frame? (iv) Resources—are the necessary resources (people, tools) available for the development?

3.1.2 Requirements Analysis

In this stage the requirements analyst:

- Identifies the stakeholders of the system: these may be customers, users, regulators, or anyone with an interest in or impacted by the system. Some stakeholders are more directly involved in the system than others—we can represent this situation by the 'onion model' of different categories of stakeholders (Fig. 3.1).
- Systematically identifies and records the requirements of stakeholders of the system, and constraints imposed on the system by existing systems it operates with, or by existing work practices and regulations.

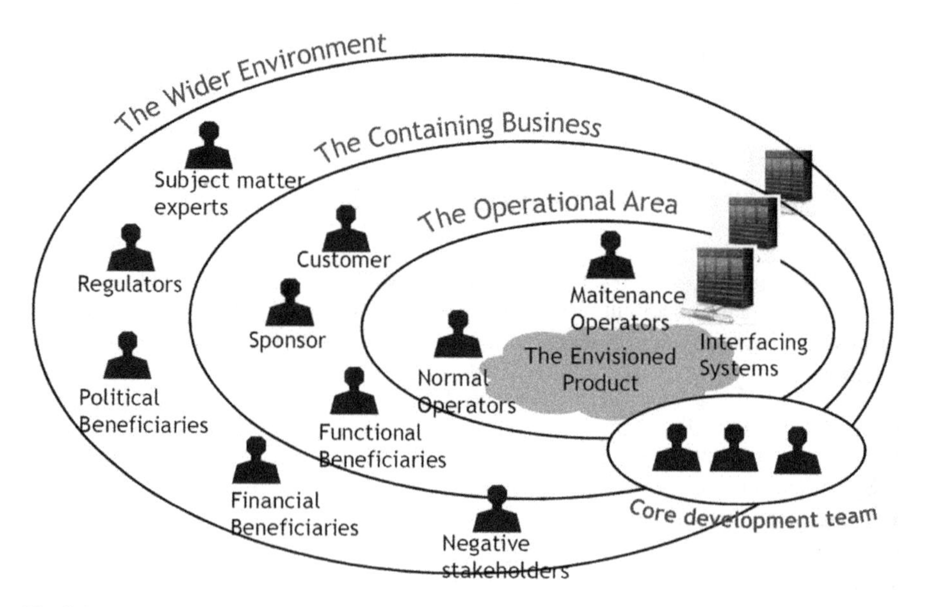

Fig. 3.1 'Onion model' of stakeholders [1]

Requirements divide into *functional requirements* (the services and functions provided by the system) and *non-functional requirements* (its efficiency, usability, extensibility, etc). There may be *conflicts* between different requirements and ambiguities in the informal requirements. Such conflicts and ambiguities should be resolved before the definitive specification is constructed.

For example, we may have an accuracy requirement for a financial computation, and a timeliness requirement, and these may be in conflict: it may be necessary to settle for lower accuracy in order to achieve the timeliness requirement, if this is the more critical requirement.

There are four main phases in requirements analysis:

- *Domain analysis and requirements elicitation*: identify stakeholders, gather information on domain and requirements from users, customers, and from other stakeholders and sources. Classify requirements as functional or non-functional, and decompose requirements into sub-requirements using a notation such as SysML [2].
- *Evaluation and negotiation*: identify conflicts, imprecision, omissions, and redundancies in requirements; consult and negotiate with stakeholders to agree resolutions to these issues.
- *Specification and documentation*: systematically document the requirements as a system specification, in a precise notation (which is not necessarily machine-readable): this specification represents an agreement between developers and stakeholders on what will be delivered.
- *Validation and verification*: check the formalised requirements for consistency, completeness and correctness wrt stakeholder requirements.

For example, in the case of a financial system the initial requirements specification could be expressed in terms of formulae defining the required properties in mathematical notation.

There are many possible requirements elicitation techniques that can be used to obtain requirements from stakeholders:

- Interviews with stakeholders
- Brainstorming sessions
- Observation of existing processes/work practices
- Scenario analysis—model specific scenarios of use of the system, e.g., as UML sequence diagrams
- Document mining
- Goal decomposition
- Exploratory prototyping.

Thorough requirements analysis can reduce errors and costs later in a development.

3.1.3 Specification

Based on the requirements specification, a detailed machine-readable model of a system is constructed, using graphical or textual notations to define required data and behaviour in an explicit but platform-independent manner. UML notations such as class diagrams and OCL can be used to define the model. It is important to avoid implementation details, in order that the specification can be translated into a wide range of different implementation platforms. Only sufficient detail should be included to specify the logical properties of the system.

Validation and verification techniques include inspection of the specification text, including formal reviews, structured walkthroughs of its behaviour in particular scenarios, and execution/testing, in the case of an executable specification language.

3.1.4 Design

Based on the specification, the design defines an architecture and structure for the system, dividing it into subsystems/modules responsible for parts of the system functionality and data. The design process includes:

1. *Architectural design*: define the global architecture of system, as a set of major subsystems, and the dependencies between them.
2. *Subsystem design*: decompose the global subsystems into smaller subsystems. Continue until clearly identified *modules* are obtained (subsystems which cannot be further divided).
3. *Module design*: define each module, in terms of:
    a. the data it encapsulates—attributes/associations;
    b. the operations it provides (external services)—e.g.: their names, input and output data, and specifications. This is called the *interface* of the module.
4. *Detailed design*: for each operation of a module, identify the steps of its processing.

The structure of a design may evolve as experience with prototypes of the system grows, for example.

3.1.5 Implementation

In this stage code is produced from the design in one or more programming languages. Traditionally, this is done manually by programmers. In model-driven engineering (MDE)/model-based development (MBD) approaches, code production is automated. Automated coding potentially reduces implementation time and cost, but problems can arise if the generated code needs to be manually adapted. The code-generation process should preserve the semantics of the specification, so that the code is *correct-by-construction* (assuming that the specification is correct). Likewise, the code generation should not introduce additional quality flaws such as duplicated code or excessive numbers of operations in a class.

3.1.6 Testing

The aim of testing is to discover errors in a system (before the customers or users discover them). Testing can be either *white-box*: based on the internal code structure, and designed to test each program path; or *black-box*: based on requirements independently of code structure.

Testing can be applied at several levels:

- Code/Unit testing: testing of each component separately (this is mainly white-box)
- Integration testing: test that components interact correctly
- System testing: test entire system
- Acceptance tests: test the system against requirements (mainly black-box).

3.1.7 Maintenance

This includes all post-delivery activities, including:

- Correction: bug-fixing and correcting defects.
- Adaption: changing the system to operate in a new or updated environment.
- Prevention: refactoring and reorganisation of a system to improve the structure of the system and facilitate its future evolution.
- Enhancement: extension of the system to handle new requirements.
- Decommissioning: ensure use of the system is phased-out in a controlled manner, and securely dispose of data held by the system.

These activities can consume far more resources than the development of entirely new systems.

3.2 Software Modelling Using UML

The Unified Modelling Language (UML), was introduced in 1997 as a unification of different object-oriented modelling approaches and methods such as OMT and the Booch method. By this time, the large number of different modelling notations was becoming an obstacle to the use of object-oriented methods, and the leading experts and companies in the field agreed to collaborate to produce a single authoritative approach. The UML is now an international standard controlled by the OMG industry consortium (including most leading software companies such as IBM, Microsoft, Oracle, etc): www.omg.org/uml. UML has become the most widely-used modelling notation in industry, and many hundreds of tools and books have been produced for UML. The language has been through two main versions 1.* and 2.*, of which version 2.* is a major extension of 1.* with richer modelling notations. Specialised versions of UML, such as Foundational UML (fUML) have also been developed, to support executable modelling or modelling in specialised domains.

The main motivations for UML modelling are:

- to *precisely define the requirements of a software application*, in a System Requirements Specification (SRS);
- *as a tool for thought*, to work towards an understanding of a system and to facilitate communication between developers and stakeholders;
- to *define reusable models* capturing domain concepts;
- to *support model-based development*, including code generation from models.

The main notations of UML are: class diagrams, use case diagrams, state machines, sequence diagrams, Object Constraint Language (OCL), activities, and deployment diagrams. Surveys of users have shown that these notations are the most frequently used, and particularly class diagrams, use case diagrams and sequence diagrams.

Class diagrams are the primary notation of UML, and define the data of an application: information that it processes or is aware of, and the structure of this data and its internal relationships. Figure 3.2 shows an example class diagram.

Other UML notations include:

- *Use case diagrams*: these describe the functionalities of an application from the user perspective. They show the functional services provided, and link these to the actors (roles of users) who may interact with the services.
- *State machines*: these define the life histories of objects in terms of states and events and the transitions that events produce between the states of objects. They can also be used to define the stages of execution of an operation or of use cases.
- *Sequence diagrams*: these show examples of interactions (communications) between objects, and between users and the system.
- *Object Constraint Language (OCL)*: a textual specification notation that can be used to precisely define operations and use cases, and class invariants.
- *Activity diagrams/textual activity notation*: these show the compositions of actions as sequences, choices, iterations, etc., to form operation or use case behaviours.
- *Deployment diagrams*: these show physical configurations of devices, communication links, and the distribution of software across devices.

Different models are emphasised in different domains. Interaction diagrams are used particularly in telecoms system specification, and for embedded systems. Activities can show workflows for business processes. Class diagrams and use case diagrams are widely used across many application domains. In the financial domain, class

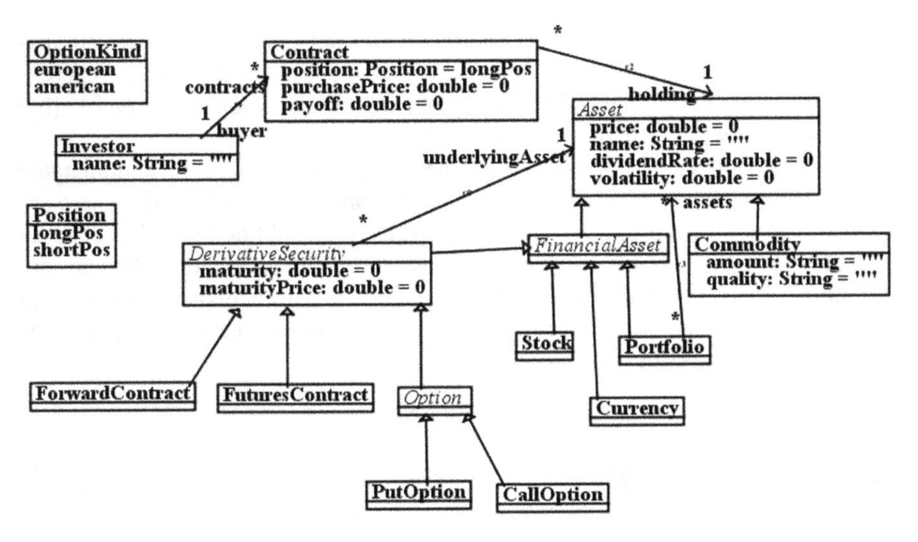

Fig. 3.2 Example UML class diagram

diagrams can be used to specify the financial concepts and their inter-relations, as in Fig. 3.2. OCL can be used to give precise functional specifications of operations, corresponding to function definitions in Excel/VBA. Activities can show the algorithms and procedures used for particular financial processes.

3.2.1 Class Diagrams

Figure 3.2 shows an example of a class diagram, for a conceptual model of derivative securities.

Class diagrams show the entity types of system as class rectangles (e.g., *Investor*) enclosing their internal data, or attributes. For example, each *Investor* has a name, which is a string, and each *Asset* has a price, name, dividendRate and volatility. This defines the structure of the application data, which can then be relied upon by application processing (e.g., the asset data can be used in calculations of the price of a derivative security based upon the asset). Relationships are shown as lines between classes: for example, any *DerivativeSecurity* has a specific *underlyingAsset*, which is an *Asset*. Relationships (associations) normally have two ends, each end may have a rolename to indicate the meaning of that end, and a multiplicity to indicate how many objects can be related: for example any number (the * symbol) of derivative securities may depend upon the same underlying asset. In addition, one investor may have any number of contracts, involving different (or the same) assets.

Specialisations between one entity and another are shown using an inheritance arrow (with an unfilled triangle at the superclass/generalisation end). For example,

Commodity is a special case of *Asset*, and *PutOption* and *CallOption* are two alternative specialisations of *Option*. Subclasses inherit all the features of their superclasses, thus a *PutOption* has attributes of *maturity*, *maturityPrice*, *volatility*, *price*, etc (but not *amount* or *quality*, which are specific to *Commodity* objects).

In cases where a small number of alternative values are needed, an enumerated type or enumeration can be used. For example, *OptionKind* declares that there are two kinds of option, european or american. This type could be used to declare an attribute *kind* : *OptionKind* of the *Option* class. Further class diagram elements are operations, which can define query or update processing on individual classes, and specialised forms of relationship, such as ordered associations and aggregations (whole/part relationships).

3.2.2 Use Case Diagrams

Use cases are used during requirements analysis of a system, to identify the ways in which the system is intended to be used, and the services it is expected to provide. A use case expresses a particular unit of an application's behaviour, encompassing a family of similar scenarios of use of the system. The users of the system are also shown, they are grouped into actors and linked to the use cases that they have the capability/authorisation to perform.

For example, for a derivative securities trading application based on Fig. 3.2, there could be use cases to buy a futures contract, to sell a futures contract and to buy and sell call options (Fig. 3.3). The actors involved in the use cases are shown as stick figures linked to their use cases. Each use case in this example involves interaction with a trader who wants to trade the asset, and with an exchange which lists these assets as tradeable.

Typically, the use cases depend upon the class diagram data and involve processing upon this data, using the data directly or by means of operations of the classes.

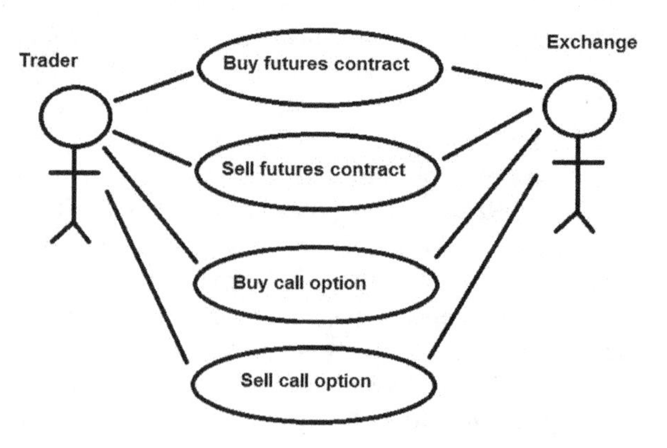

Fig. 3.3 Use case diagram example

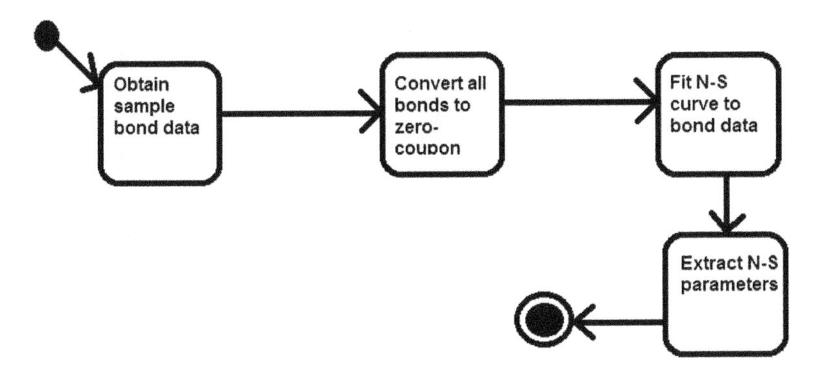

Fig. 3.4 Example state machine diagram

3.2.3 State Machines

State machines define the dynamic behaviour of objects and operations, they can also be used to specify operation or use case behaviour as a series of steps, and to express life histories of objects.

For example, the procedure to obtain a yield curve from bond market data could be specified as in Fig. 3.4. From the initial state (shown as a black circle) the process moves through states (processing stages) of obtaining market data, converting this data to zero-coupon bond data, then using an optimisation procedure to fit a Nelson–Siegal curve to the data, and finally extracting the parameters of this curve. The termination state is shown as a 'bullseye' symbol; this marks the termination of the process.

3.2.4 Interactions

Interactions, expressed as UML *sequence diagrams*, describe examples of system behaviour in terms of object communications. The diagrams show object instances *obj* of classes *Entity* as vertical lifelines, and the messages exchanged between objects as arrows between lifelines. Time increases from the top to the bottom of the diagram. Figure 3.5 shows an example interaction, in which an *Application* object *a* requests a *FIXEngine* object *f* to send an order request to set up a futures contract.

The sequence diagram shows an example (a scenario) of execution of the *buy futures contract* use case of Fig. 3.3, and details of how this use case is carried out within the system. The vertical lines show the timelines of particular objects, with time increasing from the top to the bottom of the lines. Boxes on the lines represent operation executions. An arrow with an open arrowhead, as in the example, represents an asynchronous message from one object to another: the caller does not wait for the called object to respond. An arrow with a filled arrowhead represents

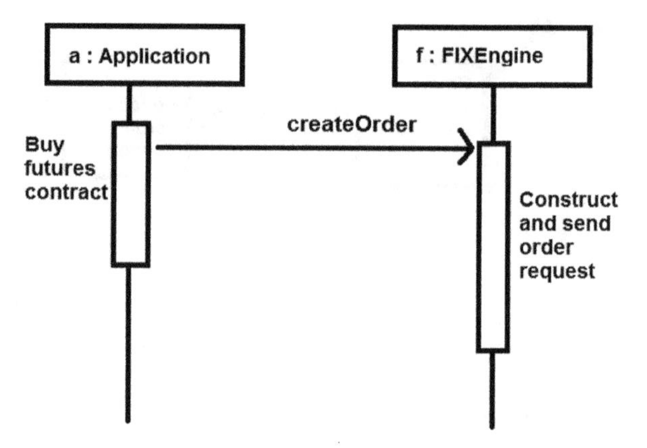

Fig. 3.5 Example interaction diagram

a synchronous message: the caller's flow of control is suspended until the called operation is completed.

Unlike class diagrams and formally-specified use case diagrams, interaction diagrams do not provide sufficient information to support code generation from the models, they are instead used to help the developers understand the required behaviour of the system. In our development process they are an optional model.

3.2.5 Object Constraint Language (OCL)

The OCL was added to UML in order to provide more precise textual specifications to describe UML elements such as operations, classes and use cases. OCL includes data types of integers and real numbers, and collection data such as sequences and sets. The usual programming language expressions such as $a = b$, $a < b$, $a \: / = b$, $a * b$, $a.pow(b)$, etc are supported, in addition various operators on objects and collections are defined, which provide a powerful facility to specify behaviours declaratively:

- Navigation through a class diagram using the $x.f$ operator to refer to feature f of object or collection x.
  For example, in the model of Fig. 3.2, *inv.contracts* is the set of contracts of an investor $inv : Investor$, and $d.underlyingAsset.price$ is the price of the underlying asset of a derivative security d.
- $s \rightarrow collect(x \mid e)$ is the sequence of values of e for objects x in a collection s.
  For example, $inv.contracts \rightarrow collect(c \mid c.holding.price)$ is the sequence of prices of the assets held via the contracts of investor *inv*.
- $s \rightarrow select(x \mid P)$ is the subcollection of collection s that consists of the $x : s$ that satisfy P.
- $s \rightarrow reject(x \mid P)$ is the subcollection of collection s that consists of the $x : s$ that do not satisfy P.
- $s \rightarrow sum()$ is the sum of the values in a non-empty collection s of numbers or strings.

- $s \rightarrow prd()$ is the product of the values in a non-empty collection s of numbers.
- $s \rightarrow sortedBy(e)$ produces a copy of s sorted in ascending order of the e-values of its elements.

These and other operators correspond to programming library operations, as in the C++ Standard Template Library (STL): $s \rightarrow sum()$ corresponds to std::accumulate (s.begin(), s.end(), 0) in C++ STL. $s \rightarrow collect(e)$ corresponds to std::transform (s.begin(), s.end(), result.begin(), e).

The operators can be combined to define the effect of operations in a class diagram, or of use cases. For example, an operation to compute a sum of squared values could be:

```
sumSquares(s : Sequence(double)) : double
post:
  result = s->collect( x | x*x )->sum()
```

3.2.6 Activity Diagrams

Activity diagrams provide a means to describe behaviours that are composed of collections of tasks (such as the algorithms of operations, or the workflows of business processes), in a graphical manner. They consist of:

Activities An activity is the specification of behaviour as the coordinated sequencing of subordinate units whose individual elements are actions.

Actions An action represents a single step within an activity, that is, one that is not further decomposed within the activity. An action may be complex in its effect and not atomic.

Activities are generalisations of sequential programming constructs such as sequencing, conditionals and loops.

Activity diagrams show:

- Actions (as state boxes)
- Arrowed lines denoting control flows (sequencing of actions)
- Conditional choice point branching and joining (diamonds)
- Parallel flows (starting and ending at vertical bars).

Figure 3.6 summarises these notations. Parallel flows mean that multiple separate threads of control execute together, whilst in a conditional choice only one flow of control is taken.

Structured activities can also be expressed as pseudocode using generic program statements:

- *variable* := *expression*
- if E then $S1$ else $S2$
- while E do S

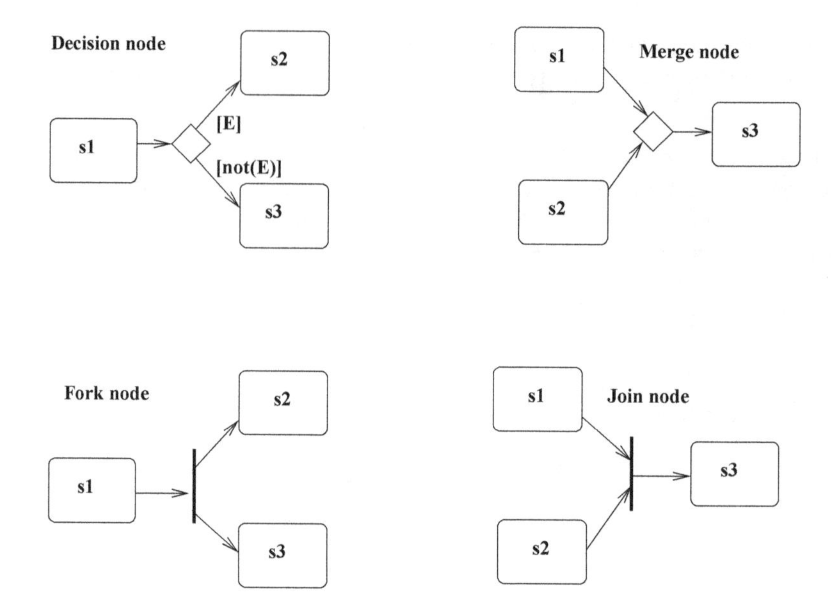

Fig. 3.6 Activity diagram notations

- $S1$; $S2$
- for x : s do S
- $S1$ parallel $S2$
- var x : T
- execute E

where S, $S1$ and $S2$ are statements, and E is an expression.

For example, the create yield curve workflow can be written more explicitly as the pseudocode statements:

```
var bonds : Sequence(Bond) ;
bonds := obtainSampleBonds() ;
var zbonds : Sequence(Bond) ;
zbonds := convertBondsToZerocoupon(bonds) ;
fitNScurveToBonds(zbonds) ;
extractNSparameters()
```

Activities are the central modelling notation for behaviours in executable UML languages such as fUML [3]. However, activities and pseudocode can be too procedural and low-level to be used for specification, and it is preferable to use OCL constraints for the declarative description of a system where possible.

3.3 Model-Based Development (MBD)

MBD is a form of software development which uses UML or another modelling language to develop systems based on models, instead of code. In MDD (Model-driven development), and MDA (Model-driven architecture) models are the primary artifact in software development, with code as secondary. Typically, executable code is automatically generated from models. MBD has the benefits that models are simpler to review and modify than code, and automated code generation can raise productivity substantially. Model repositories and libraries can be established for the rapid production of systems in a common 'product line' family. MBD may also reduce the need for outsourcing. MBD has been widely used in specialised industry sectors (such as automobile systems), with generally positive results [4]. Typically a *domain specific language* (DSL) is defined for some specific application or technical domain, and used to write models for that domain. Tools specialised for the DSL can then generate code from the DSL models, or perform analysis of the DSL models. Negative aspects of MBD are the training and adoption costs needed, and the generally poor level of tool support.

3.3.1 *Models and Metamodels*

Class diagrams and other visual and textual descriptions of a system are termed *models* of the system; they define the required and expected properties of the system using the graphical/textual notations. Such models are at a higher abstraction level than program code (in, for example, Java or ANSI C): they abstract away details of how data is arranged in computer memory or how iterations through data collections are to be performed.

At the specification level an expression such as

$$inv.contracts{\rightarrow}collect(c \mid c.holding.price)$$

can be written, without any details of how this sequence of values is to be computed or represented in a program. In a Java implementation for example, any sequential data structure such as a Vector or ArrayList could be used to implement the expression. The specification avoids fixing such a particular data structure representation, and hence permits a wide range of possible implementations in many different programming languages.

Models such as Fig. 3.2 correspond to programs in Java, C#, Python, etc: the class diagram classes will usually be represented by classes in these languages (or by `structs` in C), their features by fields/instance variables, and inheritance by an inheritance mechanism in these languages.

Class diagrams can also describe *languages* themselves, e.g., a language of activities (Fig. 3.7). Such diagrams are termed *metamodels*: they are models defining the

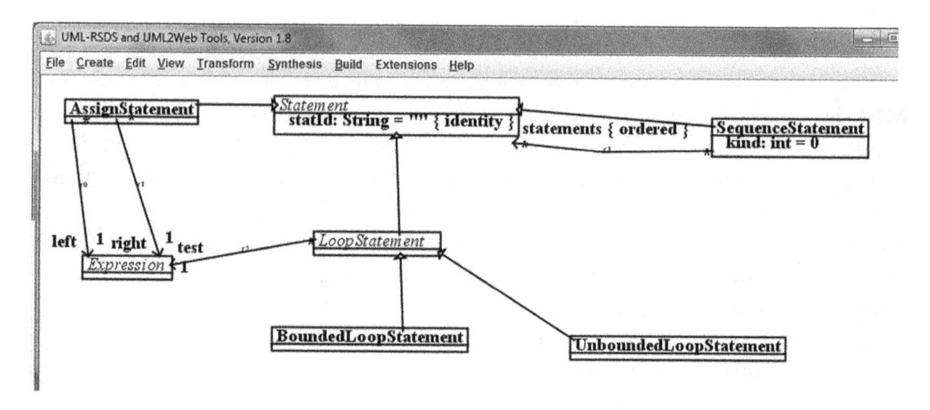

Fig. 3.7 Activity language as a metamodel

permitted structure of other models, in this case of the activity language statements. Metamodels can be defined to represent software languages such as Java, C, Python etc, as well as the entirety of UML itself.

3.3.2 Model Transformations

Model transformations are procedures which map one model into another model, or into text. An example is a *code generator* transformation which takes as input a class diagram model, and produces as output Java code. Other forms of transformation include *refactorings*, to restructure a model in-place, or *migrations*, to map a model in one modelling language to a corresponding model in a related modelling language. In this book we will be mainly concerned with the use of transformations, rather than their definition, but the UML notations of class diagrams (used as metamodels) and use cases can also be used to specify model transformations in addition to conventional software applications.

3.3.3 Domain-Specific Modelling

A *domain-specific language (DSL)* is a notation (graphical, textual or a combination), together with a precise definition of the notation, intended to represent concepts and elements of specific application domains. A DSL has:

- An identified application domain where it will be used.
- A set of concepts with properties and relationships, forming the *ontology* of the domain.

- Representations that are appropriate, visually and conceptually, for modellers in the domain, and for stakeholders who need to review such models. This is termed the *concrete syntax* of the DSL.
- A precise abstract grammar, defining the way DSL elements can be combined. This is called the *abstract syntax* of the DSL. It can be defined by a metamodel, e.g., Fig. 3.2 considered as a metamodel. This fixes the structure of models of the DSL.
- A precise way, usually by means of model transformations or templates, to map DSL models to implementations.

DSLs encode domain knowledge about a specific technical or application domain, and the DSL transformations encode knowledge and expertise about the implementation of the domain concepts in code. Domains are typically quite narrow, such as a domain for derivative security pricing. The restriction of the domain enables more focussed and accurate modelling, and specifically optimised code to be generated, however this also means that multiple DSLs may need to be constructed for a single application.

3.3.4 The State of Practice of MBD in Industry

The research of [4] surveyed over 450 practitioners, and found that some MBD is used across many sectors, including finance. But it is usually used selectively, for parts of systems. Commonly, a DSL is built for a narrow domain (a family of closely-related applications), together with code-generators and templates for this domain. This automates the previously manual coding of applications in the domain. Productivity increases of 20–30% were obtained from code generation. It was concluded that MBD can reduce effort in developing and maintaining software—enabling businesses to concentrate on their core business, not IT. Numerous companies reported that they reduced offshoring, as MBD automated work that would previously have been outsourced.

The survey also found that there were barriers to MBD adoption, particularly from middle-managers and from code gurus (who may oppose MBD because of their fear of their skills becoming redundant). To be successful, MBD projects need at least one 'MBD guru' in a development team. MBD seems more appropriate for specific domains, not for general purpose software.

Some examples of substantial MBD projects include:

- SunGard Financial Systems carried out a MBD modernisation project of Front Arena/AMS (an application concerned with the management of trades and orders). This project used MDA and Scrum—but with XML, not UML [5].
- Motorola use the MDD-SLAP process for telecomms systems development [6].
- Tata Consultancy Services defined the MasterCraft toolset for business systems development, using MBD and UML [7].

- Volvo Cars use a systematic MBD process for vehicle software construction, using Simulink models of vehicle components [8].

Companies such as Volvo and Tata have invested heavily in MBD and have used it in many projects. Microsoft have used an MBD approach called Software Factories, for software product line development using DSLs [9].

3.3.5 MBD Using UML-RSDS

UML-RSDS is a lightweight MBD approach, in which platform-independent specifications are defined using UML class diagrams and use cases. These are analysed for internal quality and correctness. Platform-independent designs are then synthesised in a mainly automated manner from the specifications. From the designs, executable code in an OO programming language (currently, Java, C# or C++) or in C or Python can then be mainly automatically synthesised.

This approach is more restricted in scope than more elaborate MBD approaches such as MDA, but it is more automated (Fig. 3.8). Thus it provides greater agility and ability to respond to changes. In contrast to fUML, applications are defined in terms of declarative use case and operation specifications, instead of detailed procedural activities. Thus specification construction and modification is simpler and less costly.

DSLs can be defined in UML-RSDS using class diagram metamodels to define the domain ontology and DSL abstract syntax. A simple concrete syntax is used, based on OCL, and transformations defined using UML-RSDS can be used to map DSL models to code or to other representations.

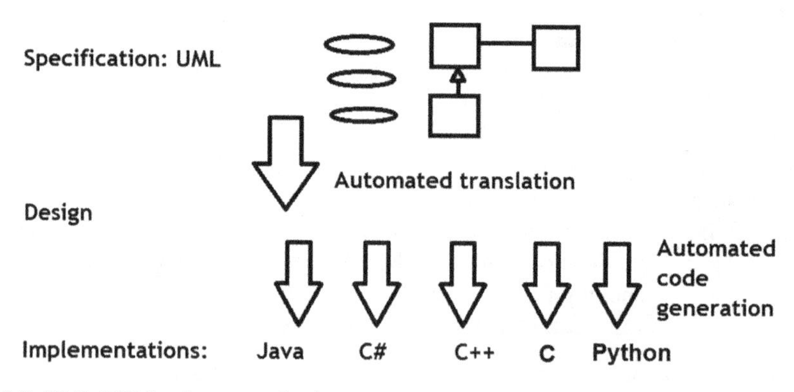

Fig. 3.8 UML-RSDS software production process

3.4 Agile Development Methods

The concept of agile software development was introduced c. 2000 to correct drawbacks of conventional plan-based development of software, which was viewed as a heavyweight and inflexible process, unable to respond to changing requirements [10]. Conventional plan-based development emphasises the gathering and formalising of all requirements before starting coding, in contrast, agile development emphasises incremental work on requirements, coding, and integration in short cycles. Thus it can be responsive to changing requirements.

In conventional development, the development team is often isolated from stakeholders: this results in delays in obtaining information or feedback. In contrast, agile development emphasises close collaboration between the team and stakeholders.

The key principles of agile development include (agilemanifesto.org): to satisfy the customer through early and continuous software delivery; to welcome changing requirements; to deliver working software frequently (every 2 weeks to every 2 months); business people and developers to work together daily; rely on face-to-face communication to convey information; continuous attention to software quality; simplicity is essential.

Agile development is now widely adopted in industry. The main approaches are eXtreme Programming (XP) [10, 11], Kanban and Scrum [12].

3.4.1 Agile Development Techniques

Agile techniques include:

- Iterations or *sprints*: development work which implements specific user requirements, in a short time frame to produce new releases.
- *Refactoring*: Regularly restructure code to improve it, to remove redundancy and other flaws [13].

Figure 3.9 shows the agile development process used in the Scrum method, with a *product backlog* of work items to be worked on in the project, a subset selected for work in the current iteration, the *sprint backlog*, and an iteration or *sprint* which implements these work items to produce a deliverable increment of the software. The iteration should not normally last longer than one month, and a daily review cycle is maintained via 'daily scums' or 'standup meetings' of the team.

Sprints are regular re-occuring iterations in which project work is completed. These produce deliverables that contribute to the overall project and yield an increment of the system. Each iteration involves a set of work items or tasks ('user stories' in Scrum) to be implemented. Tasks can be classified by their business value to the customer (high, medium, low), and by the development risk or the development effort. High priority and high risk tasks should be dealt with first. The project *velocity* is the amount of developer-time available per iteration. Taking these factors into

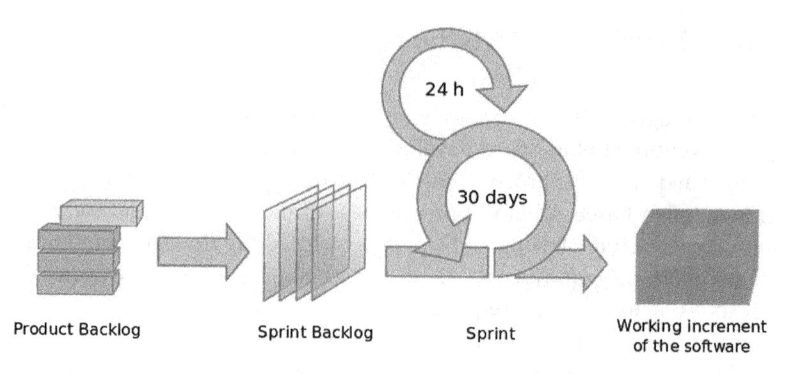

Fig. 3.9 Scrum process

account, it is possible to define an initial *release plan*: identifying which tasks will be delivered by which iteration and by which developers. This plan will be revised as development proceeds.

3.4.2 Agile Methods: Scrum

The Scrum method is now the most widely-used agile approach. It involves the following key elements:

- *User stories*: requirements expressed in terms of capabilities needed by users of the system in specific roles. The general template for a user story is

  "As a [user role], I want to [goal], so that I can [achieve business value]"

  E.g., "As a customer of the bank, I want to view the balances of my bank accounts, so that I can manage my finances".
  User stories are essentially use cases, and can be decomposed into subtasks/subcases.
- *Product Backlog*: an ordered (in terms of priority) list of user stories/tasks relevant to the project.
- *Sprint Backlog*: an ordered list of user stories/tasks to be completed in a sprint.
- *Sprint planning*: performed by the Scrum team before a sprint, the team agrees the tasks to be worked on in the sprint, that is, the subset of the project backlog to include in the current sprint backlog. The duration of this meeting is normally 2 h per each week of the sprint.
- *Daily scrum*: this daily meeting organises the activities of the team, reviews sprint progress, and identifies issues. It is time limited (e.g., 15 min) and often takes place at the start of each day. It raises the key questions for developers: (i) what did I achieve yesterday? (ii) what do I plan to achieve today? (iii) is there anything blocking me from achieving my work? The meeting is also called the 'daily

Table 3.1 Schematic Scrum board structure

Product backlog	Sprint backlog	Under development	Completed	Delivered
Task15	Task9	Task7	Task3	Task1
Task16	Task10	Task8	Task4	Task2
Task17	Task11		Task5	
Task18	Task12		Task6	
Task19	Task13			
Task20	Task14			

standup' meeting. It is not for detailed discussion, problems identified should be dealt with by a designated person following the meeting.

- *Definition of done*: a criteria to state when a task is completed. For example, that the testing process for it has been completed and all bugs detected have been fixed.
- *Sprint review*: a review conducted by the team at the end of the sprint. It identifies what work was completed and planned work that was not completed, demonstrates completed work to the stakeholders, and involves collaboration with the stakeholders to decide what should be in the following programme of work—for example in which iteration remaining uncompleted tasks should be completed. A review is recommended to last 1 hour for each week of the sprint it reviews.
- *Sprint retrospective*: after the sprint review, before the next sprint planning. It analyses the achievements of the sprint, and considers ideas for improvement of the development process. It is facilitated by the Scrum master and has a duration of approximately 1.5 h for each 2 weeks of the sprint.

Additional optional elements, such as an *impediment backlog* can also be included.

During a sprint, the team uses a *Scrum board* showing the tasks to do, in progress and completed (for example, Table 3.1).

A *Burndown Chart* shows a graph of the estimated remaining work against time.

The key roles of the Scrum team members are:

- **Product owner**: this team member is a stakeholder representative in the team, and is responsible for liaising between the technical staff and the stakeholders. The product owner identifies required work items, identifies their priority, and adds these to the product backlog.
- **Development team**: the workers who perform the technical work. The team should have all needed skills and be self-organising. Typically there will be between 3 to 9 team members in this role.
- **Scrum master**: the Scrum master facilitates the Scrum process and events, and the self-organisation of the team. This role is not a project manager role and does not have personnel management responsibilities. They ensure that the Scrum process is correctly followed in the development.

3.4.3 Agile Methods: Extreme Programming (XP)

XP is an alternative agile approach, with a strong focus on the coding activity, in contrast to Scrum, which emphasises the organisation of the development team and process.

XP advocates techniques such as *pair programming*, where two programmers work at one terminal, one having the role of reviewing the code of the other. Code refactoring is also a key process in XP, to achieve the agile principle of "continuous attention to software quality".

XP consists of:

- 5 Values: *communication*, *simplicity*, *feedback*, *courage*, *respect*.
- 3 Principles: *Feedback*—via customer interaction and fine-grained testing; *Assuming simplicity*—code that's just good enough, small changes; *Embracing change*.
- 12 Practices: *pair programming*; *planning game*; *test-driven development*; *whole team*; *refactoring*; *small releases*; *system metaphor*; *simple design*; *continuous integration*; *collective code ownership*; *sustainable pace*; *coding standards*.

XP has been found more appropriate for small teams, including single-programmer developments.

3.4.4 Agile Methods: Kanban

Kanban is a general manufacturing concept originating in the Japanese car industry: the principle is only to produce what is needed when it is needed, not before; demand is tracked through all production stages in order to manufacture to meet demand. Adapted to software, it is an agile approach oriented to a continuous delivery software production process. The approach uses demand-led production, the demand (such as customer or user requests) determines the priority for work items.

An important principle is to limit the work-in-progress (WIP): developers work only on few tasks—often just one—at a time. When the current task is finished, they then start on the next highest priority task from the backlog.

Table 3.2 compares Kanban and Scrum.

Kanban uses *Kanban boards*, which are similar to Scrum boards, and may have columns such as *Backlog* (tasks to be done next); *In Development*; *Testing*; *Customer Acceptance*; *Done*.

With Kanban it is possible to have separate teams for separate development stages (e.g., a specification and design team, and a coding and delivery team). The output of one team is fed into the backlog of the next.

Table 3.2 Kanban compared to Scrum

Kanban	Scrum
No prescribed roles	Roles of Scrum master; team member; product owner
Continuous delivery	Timeboxed sprints
Work pulled through system (single task flow)	Work performed in batches (sprint backlogs)
Environments with high variation in task priorities	Environments where tasks can be batched and worked on together

3.4.5 Benefits and Disadvantages of Agile Development

The State of Agile Survey of the software industry (versionone.com, 2017) has consistently identified that the main benefits of agile development are: (i) the ability to manage changing priorities; (ii) increased team productivity; (iii) improved project visibility.

The majority of survey respondents found that agile projects were mainly successful. Scrum is consistently the main method used, with iteration planning, daily standups, retrospectives, reviews, short iterations, release planning the most common techniques.

The following disadvantages of agile development have been identified:

- That it is focussed on manual coding—and hence is resource intensive.
- It is focussed on functional requirements and does not explicitly address non-functional requirements.
- It does not emphasise reuse.

Generally, agile approaches minimise the use of documentation and formal architecture descriptions, and verification is carried out only by testing or by code inspection.

The state of agile survey found that barriers to the adoption of Agile development included:

- Organisational culture in conflict with agile principles.
- Lack of experience with agile methods.
- Lack of management support.
- Lack of access to users/customers.

3.4.6 Can Agile and MBD be Combined?

Both agile and MBD aim to accelerate development. Whilst agile focusses upon rapid response to changing requirements, MBD focusses upon software correctness and adding value over a longer term. In principle, the combination of agile and MBD approaches could be effective and provide complementary strengths:

- MBD provides modelling, verification and reuse support lacking in Agile.
- An agile MBD approach would use models as the primary artifact, not code, reducing development costs and time.

But MBD tends to be a 'heavyweight' process with substantial use of documentation and multiple process steps. In particular, the use of multiple models in UML and MDA hinder rapid specification change, since a change to one model may impact others, and these inconsistencies need to be resolved before a new executable version is generated. The agile emphasis on simplicity can be applied to reduce the complexity of MBD, in particular to reduce the number of parallel models being maintained for a system.

Several agile MBD approaches have been created [14, 15], including MDD-SLAP at Motorola [6] and xUML/fUML/Alf from the OMG [3]. In this book we will follow an agile/MBD process based on the UML-RSDS subset of UML and its supporting tools.

Summary

In this chapter, we:

- Introduced the concepts of UML, MBD, DSLs and Agile development
- Gave examples of their use in industry, and rationales for/against their use.

In subsequent chapters, we will use an agile MBD approach for developing financial systems, based on UML.

References

1. I. Alexander, Stakeholders - who is your system for? Comput. Control. Eng. **14**(1), 22–26 (2003)
2. OMG System Modeling Language Specification Version 1.5 (2008), www.omg.org/specs/SysML
3. OMG, Semantics of a Foundational Subset for Executable UML Model (FUML), v1.1 (2015)
4. J. Whittle, J. Hutchinson, M. Roucefield, The state of practice in Model-driven Engineering. IEEE Softw. 79–85 (2014)
5. M.B. Nakicenovic, An agile driven architecture modernization to a model-driven development solution. Int. J. Adv. Softw. **5**(3, 4), 308–322 (2012)
6. Y. Zhang, S. Patel, Agile model-driven development in practice. IEEE Softw. **28**(2), 84–91 (2011)
7. V. Kulkarni, S. Barat, U. Ramteerhkar, Early experiences with agile methodology in a model-driven approach, *MODELS*, (Springer, Berlin 2011), pp. 578–590
8. U. Eliasson, R. Heldal, J. Lantz, C. Berger, Agile MDE in mechatronic systems – an industrial case study, *MODELS* (2014)
9. J. Greenfield, K. Short, S. Cook, S. Kent, *Software Factories* (Wiley, New York, 2004)
10. K. Beck et al., *Principles Behind the Agile Manifesto*, (Agile Alliance, 2001), http://agilemanifesto.org/principles

11. K. Beck, C. Andres, *Extreme Programming Explained: Embrace Change*, 2nd edn. (Addison Wesley, 2004)
12. K. Schwaber, M. Beedble, *Agile software development with Scrum*, (Pearson, 2012)
13. M. Fowler, K. Beck, J. Brant, W. Opdyke, D. Roberts, *Refactoring: Improving the Design of Existing Code* (Addison-Wesley, 1999)
14. S. Hansson, Y. Zhao, H. Burden, How MAD are we?: Empirical evidence for model-driven agile development (2014)
15. G. Guta, W. Schreiner, D. Draheim, *A lightweight MDSD process applied in small projects*, in *Proceedings 35th Euromicro conference on Software Engineering and Advanced Applications*, (IEEE, 2009)

Chapter 4
Financial System Specification Using UML

In this chapter we will explain how UML models can be used to specify financial applications, using examples.

We will cover:

- Class diagrams: classes, attributes, associations, inheritance, operations.
- Use case models.
- OCL (Object Constraint Language).
- Specification revision and refactoring.

4.1 Class Diagrams

Class diagrams show the *entity types* of a system: the data types (classes) which have instances (objects) with identities and internal structure. For example, Customer, Account, etc (Figs. 4.1, 4.4).

Value types include integers, reals, strings and booleans. These do not have instances with identity or internal structure, but are simply values (although strings can also be considered to be structured as sequences of characters).

Collection types include sets and sequences: sets are collections of elements which have no ordering and no duplicates—it is only possible to ask whether an element x is in the set s ($x : s$) or not in the set ($x \notin s$). Sequences sq have an order and allow duplicates. Thus in addition to membership, we can obtain the ith element of a sequence ($sq[i]$) and find out how many times an element x occurs in sq ($sq \rightarrow count(x)$).

K. Lano and H. Haughton, *Financial Software Engineering*, Undergraduate Topics in Computer Science, https://doi.org/10.1007/978-3-030-14050-2_4

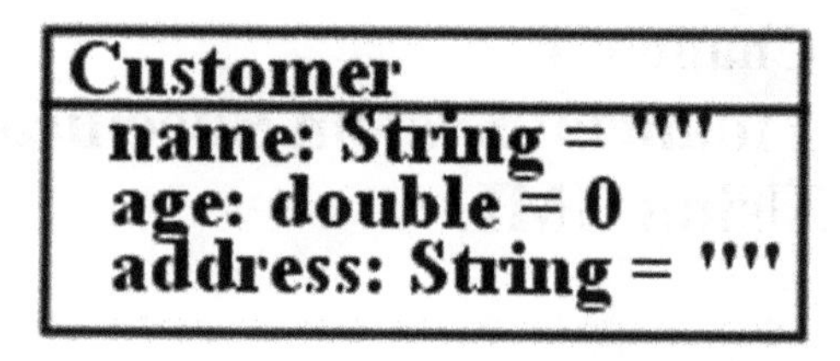

Fig. 4.1 Example of class specification

4.1.1 Classes

Classes have a name (usually singular, with an initial capital), and they have a series of *attributes* of value type. Class specifications formalise requirements such as "For each customer, their name, age and address are recorded".

Class diagrams can be used for initial conceptual modelling of a system, for system specification, informal and formal, and as an executable specification in MBD. Usually a UML class can be translated directly to a Java, C#, C++ class, or to a C `struct`. For example:

```
class Customer
{ String name = "";
  double age = 0;
  String address = "";

  ...
}
```

in Java, for the class of Fig. 4.1.

4.1.2 Attributes

Attributes represent intrinsic and permanent properties of an object (class instance). The attribute value can change over time, but must always be of the declared type. Attributes are usually written with a lowercase initial letter.

If $att : T$ is declared in class C, and obj is an instance of C, then $obj.att$ is a value of type T.

We use the computational data types `int` (32-bit signed integers), `long` (64-bit signed integers) and `double` (double-precision floating point numbers) for numeric values:

- `int` values are between -2^{31} and $2^{31}-1$ (-2147483648 to $+2147483647$).
- `long` values are between -2^{63} and $2^{63}-1$
- `double` values are assumed to satisfy the IEEE 754 floating point standard, and range from $-1.7976931348623157E+308$ to $+1.7976931348623157E+308$.

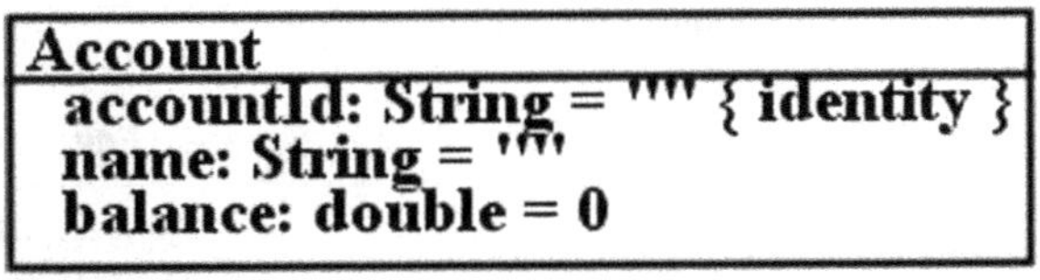

Fig. 4.2 Identity attribute example

Account
accountId: String = "" **{ identity }**
name: String = ""
balance: double = 0

These correspond to the data types available in Java, C#, C++ and C, although their sizes may vary in C and C++. Using such datatypes (instead of abstract mathematical Integer and Real types) reduces the semantic distance between specification and implementation, and simplifies the verification of implementations against a specification. The specifier must be aware of the bounded nature of the datatypes and ensure that bounds are not exceeded during computations. In addition, exact computations are not possible with doubles due to their bounded precision. The String data type corresponds to the Java String and C# string types. String values are written between double quotes as usual. Boolean values *true* and *false* are elements of the type boolean.

Attributes of collection type are also possible, for example an attribute *rates*: *Sequence*(*double*) could store a sequence of interest rates in successive time periods.

Identity attributes are special kinds of attributes *eId*: *String* which uniquely identify objects of their class (Fig. 4.2). If objects $e1$: E and $e2$: E have $e1.eId = e2.eId$, then the objects are the same ($e1 = e2$). This is related to the concept of *primary key* for relational databases, and provides a means of looking up objects based on a key value. Typically the id values are entered by a user in a user interface, and the application then looks up the actual objects identified by these values, for example, in a database.

We use the notation $E[eval]$ for the instance of E with *eId* value *eval*, where *eId* is the first identity attribute defined for E.

4.1.3 Enumerated Types

New finite value types can be introduced as enumerations in a class diagram. Distinct named values are listed in an $\ll$ *enumeration* $\gg$ rectangle (as for *AccountKind* in Fig. 4.3). The enumerated type can then be used as the type of attributes elsewhere in the diagram (e.g., *kind* : *AccountKind* in the *Account* class).

4.1.4 Associations

Associations define relationships between classes, the elements of an association are links or pairs $obj1 \mapsto obj2$ of instances of the source and target classes. There should be multiplicities at both ends, and a role name at least at one end. The role

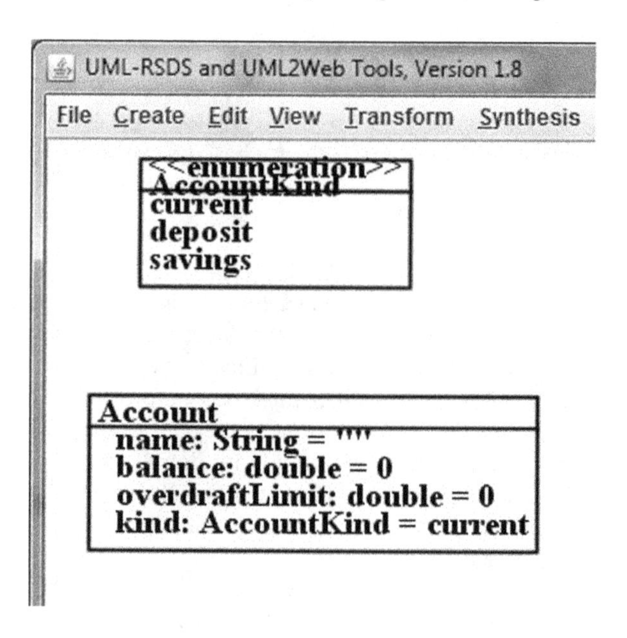

Fig. 4.3 Enumeration example

name at the target of the arrow (role2) is mandatory, role1 (at the start of the arrow) is optional. Each role name becomes a feature of the class at the *opposite* end of the association.

Associations formalise requirements such as "Each customer has a set of accounts, and each account may belong to several customers" (Fig. 4.4). The role name *accounts* represents navigation from a customer to its set of accounts, likewise *customers* is the set of customers linked to a given account. The arrow on the association indicates that navigation in that direction must be supported by an implementation.

Association multiplicities specify how many objects of one class may be linked to an object of another via the association (Table 4.1).

The * notation means the role at that end is a set or sequence of objects of the class at the end, of unspecified size. 0…1 means the role is a set/sequence of size ≤ 1. 1 means it is a specific (non-null) object of the end class. In general a multiplicity range $a \ldots b$ means that there can be between a and b number of objects in the association end.

- If $A\ {}^{m1}\!\!-\!\!-{}^{*}_{r}\ B$ for any multiplicity $m1$, then for each instance *obj* of A, *obj*.*r* is a set (possibly empty) of B objects.
- E.g., *c*.*accounts* for customer *c*.

In terms of program code a 1 or 0…1 end is usually represented as a single object, a * end is represented as a collection:

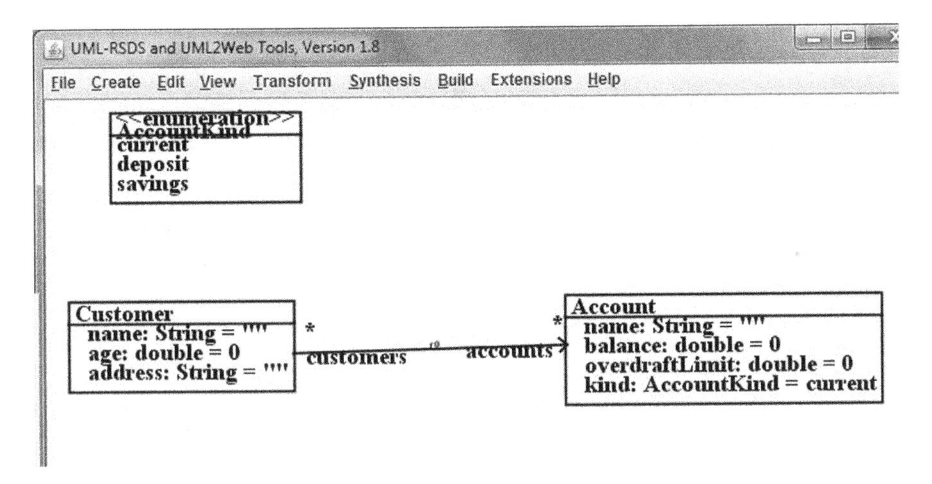

Fig. 4.4 Association example

Table 4.1 Association multiplicities

Multiplicity	Meaning
*	Any finite number of objects at this end can be linked to one object at the other end
0...1	At most one object at this end can be linked to one object at the other end
0...n	At most *n* objects at this end can be linked to an object at the other end
1	Exactly one object at this end is linked to one object at the other end
n	Exactly n objects at this end are linked to one object at the other
1...*	At least one object at this end is linked to one object at the other

```
class Customer
{ String name = "";
  double age = 0;
  String address = "";
  Set<Account> accounts = new HashSet<Account>();
  ...
}
```

- If $A \ {}^{m1}\!\!-\!\!{}^{1}_{r} \ B$ for some multiplicity $m1$, then for each instance *obj* of A, *obj.r* is a single *B* object. These are referred to as *many-one* associations, if $m1 \neq 1$.
- If $A \ {}^{m1}\!\!-\!\!{}^{*}_{r} \ ^{\{ordered\}} B$ then for each instance *obj* of A, *obj.r* is a sequence (possibly empty) of *B* objects. Individual elements of *r* can be referred to as $r \rightarrow at(i)$ or $r[i]$ for the *i*th element of *r*, with numbering starting at $i = 1$.

- If $A\ ^{m1}\!\!\stackrel{0..1}{\underset{r}{—}}\ B$ then for each instance *obj* of A, $obj.r$ is a set of 1 or 0 (empty set) of B objects, or an optional B object. We use the empty collections *Set*{} or *Sequence*{} instead of explicit *null* values for $obj.r$ at the specification level.
- One-to-one associations $A\ ^{1}\!\!\stackrel{1}{\underset{r}{—}}\ B$ are unusual, and represent a bijection between A elements and B elements.

The case of a $A\ ^{m1}\!\!\stackrel{m2}{\underset{r2}{—}}\ B$ association with neither $m1$, $m2$ being 1, is termed a *many-many* association.

Bi-directional associations have role names r1 and r2 at both ends:

- If $A\ {}^{m1}_{r1}\!\!—\!\!{}^{m2}_{r2}\ B$ then $r1$ is a feature of B, with multiplicity $m1$, and $r2$ is a feature of A, with multiplicity $m2$.
- $r1$ and $r2$ depend on each other: if the pair $a \mapsto b$ is in the association, then b is a value of $a.r2$, and a is a value of $b.r1$.
- Maintaining this mutual consistency is difficult using hand-written code
- Bi-directional associations create strong semantic links between classes; so they should only be used if they are essential to the modelling of a problem.

E.g., the *Customer—Account* relationship.

Aggregations are special kinds of association which model situations where one class has a whole-part relation to another (e.g., a bond has a number of cash flows). Aggregation is represented by a black diamond symbol at the 'whole' end. The semantic effect is that if a 'whole' object is deleted, so are all its linked parts (cascaded delete). The multiplicity must be 1 or 0...1 at the 'whole' end.

4.1.5 Inheritance

Inheritances define specialisation/generalisation relationships between entities (Fig. 4.5). The inheritance arrow points from the subclass (specialised entity type) to the superclass (generalised entity). No rolenames or multiplicities are written on the line. The superclass is usually an *abstract* class: instances cannot be created for it, only for its concrete subclasses. Abstract classes are shown by writing their name in italic font (*Customer* in Fig. 4.5).

It is possible to have *multiple subclassing*: several specialisations of the same superclass (e.g., *PersonalCustomer* and *BusinessCustomer* as subclasses of *Customer*). It's more unusual to have *multiple inheritance*: several superclasses of one subclass (e.g., *HouseBoat* as a subclass of both *Residence* and *Boat*), but this is permitted in UML. All features of all superclasses are inherited by a subclass, and operations of a subclass may override operations with the same name and input parameter types in the superclass. In this respect, UML inheritance is the same concept as Java `extends` and C++ `public` inheritance.

A *leaf class* is a class at the base of the inheritance hierarchy, with no subclasses. Such classes should be concrete (not abstract) in a completed specification. A *root class* is a class at the top of the inheritance hierarchy, with no superclasses.

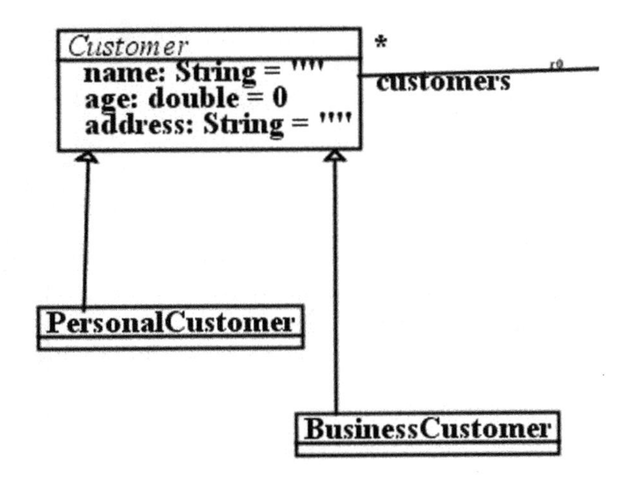

Fig. 4.5 Inheritance example

If an object is a member of a subclass *Sub*, then it is a member of every superclass (direct or indirect) *Sup* of *Sub*:

$$x : Sub \Rightarrow x : Sup$$

So, for example, if *pc* is a *PersonalCustomer*, it is also a *Customer* instance.

If abstract class A is a direct superclass of $B_1, \ldots, B_n$ (and these are the only direct subclasses of A), then if $x : A$, it must also be a member of exactly one of the B_i. Thus any *Customer* object must either be a *PersonalCustomer* instance or a *BusinessCustomer* instance.

4.1.6 Operations

Operations of a class are either (i) *query operations*: these return a value computed from the object data, and do not update the object state or (ii) *updaters*: which modify object state (and may return a value). Operations can be specified by *preconditions* and *postconditions*: expressions that define necessary assumptions at the start of execution, and that define the result state (and possibly the return value) at termination. Alternatively, behaviour can be defined by a statemachine or by an activity/pseudocode.

In Fig. 4.6 *totalFunds*() : *double* is a query operation. *withdraw*(*amt* : *double*) is an updater.

The operations can be specified by pre and post conditions:

```
query totalFunds() : double
pre: true
post:
  result = accounts->collect(balance)->sum()
```

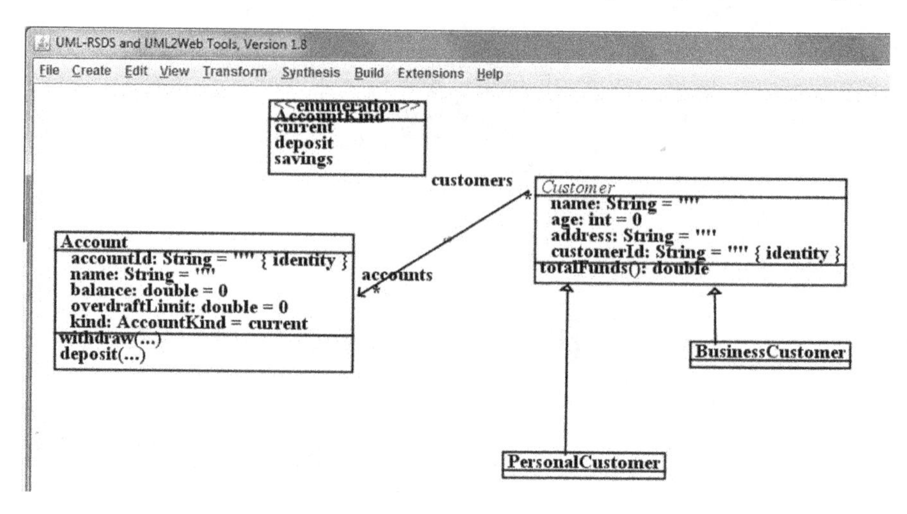

Fig. 4.6 Classes with operations

Precondition *true* means the operation is always valid to execute.

```
withdraw(amt : double)
pre: balance - amt >= -overdraftLimit
post:
  balance = balance@pre - amt
```

The *withdraw* operation should only be invoked if the condition $balance - amt \geq -overdraftLimit$ holds. The *amt* is then subtracted from *balance*. In the postcondition, *balance*@pre refers to the previous value of *balance* (the value at the time when the operation is invoked). Operations can be defined in classes or in use cases. Call-by-value semantics is used for parameters of basic data types (numerics, booleans, strings). Changes cannot be made to the parameter within the operation. For object and collection-valued parameters changes can be made to the contents of the parameter (attribute values of objects and membership of collections) and these changes will be retained after execution of the operation.

Query operations must have a result type, and some equation $result = e$ as the last conjunct in each conditional case in their postcondition. Update operations normally have no return type, but this is possible. Operations can be defined recursively, e.g.:

```
static query realfact(d : double) : double
pre: true
post:
  (d <= 1  =>  result = 1) &
  (d > 1  =>  result = d*realfact(d-1))
```

Operations can be called by the notation *obj.op(params)* as usual. Such calls are expressions, for query operations, and statements for update operations. If *op* is a query operation then the expression can be used within other expressions as a value of *op*'s return type. Update operations should not normally be used as values or in contexts (such as conditional tests or constraint antecedents) where a pure value expression is expected. Callers of an operation must ensure its precondition at the point of call—otherwise an exception may occur in the generated code. If called with its precondition satisfied, the operation then guarantees its postcondition at termination.

Precondition conditions should be sufficient to ensure that postcondition expressions are well-defined: that no division by zero or other undefined computation can occur, and that numeric computations remain within the bounds of defined numeric types.

Another example of an operation definition is the normal distribution function $N(m, \sigma)(x)$, from the *NormalDist* library:

```
static query normal(m : double, sigma : double,
                    x : double) : double
pre: sigma > 0
post:
  disp = x - m &
  denom = ( -0.5*disp*disp/(sigma*sigma) )->exp() &
  num = sigma*(2*MathLib.pi())->sqrt() &
  result = denom/num
```

Note that $sigma > 0$ is needed in the precondition, to ensure the definedness of the postcondition.

Local variables such as *disp* can be introduced (in query operations) by an equation $var = value$ and then subsequently used in the postcondition. A pre-state expression *f@pre* for a feature *f* of the owning class can be used in operation postconditions and use case postconditions. In an operation postcondition, *f@pre* is the value of *f* at the start of the operation—this is the same value as denoted by *f* in the operation precondition. Occurrences of *f* without *@pre* in the postcondition refer to the value of *f* at the end of the operation.

So, for example:

```
op()
pre: b > a
post: b = b@pre * a@pre
```

multiplies *b* by *a*. Since *a* itself is not updated, there is no need to use *pre* with *a*, and this operation should be written as:

```
op()
pre: b > a
post: b = b@pre * a
```

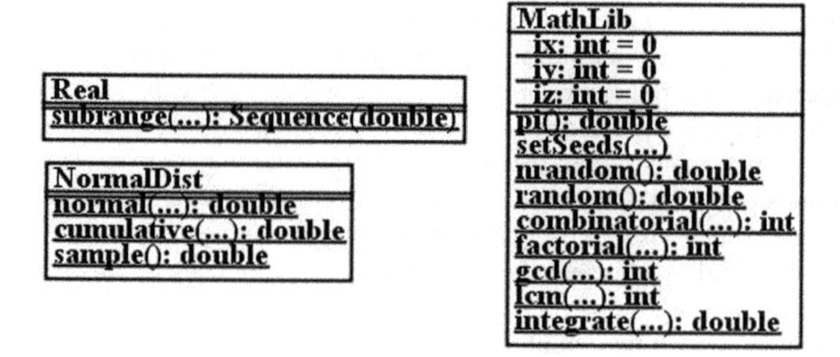

Fig. 4.7 Static operations and attributes

The keyword *static* in the above definitions of *realfact* and *normal* indicates that the operation is defined on a class, independently of particular instances. Static operations are also known as *class-scope* operations, in contrast to *instance-scope* operations. Static operations *op* of class *E* are invoked using the class name: *E.op(pars)*.

Attributes may also be static, this means that every object of the class sees the same value for the attribute. Both static operations and static attributes are underlined in class diagrams (e.g., Fig. 4.7).

Another static operation example is a random number generator, from *MathLib*:

```
static random() : double
pre: ix > 0 & iy > 0 & iz > 0
post:
  ix =  (ix@pre*171) mod 30269 &
  iy =  (iy@pre*172) mod 30307 &
  iz =  (iz@pre*170) mod 30323 &
  r = ( ix/30269.0 + iy/30307.0 + iz/30323.0 ) &
  result = r - r.floor
```

This is a static operation which updates the *ix*, *iy*, *iz* static int attributes each time it is called. These need to be initialised to suitable seed values.

4.2 Use Case Models

Use case models define the functionalities of a system, the services it provides to users. Each use case has a name, written in an oval, and linked to agents/actors who interact with the case. E.g., *checkBalance* is a use case that can be used by customers or bank staff, whilst *createAccount* is only for staff (Fig. 4.8).

Use cases are defined by a sequence of steps, each step can perform operations on objects of the system. Use cases coordinate object behaviours to produce an overall required functionality. E.g., *checkBalance(aId : String)* has these steps:

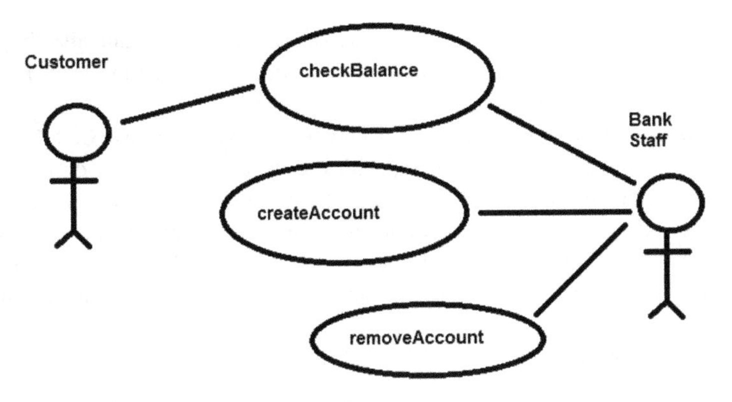

Fig. 4.8 Use case model example

```
Lookup account with accountId = aId;
Display balance of this account;
```

createAccount(cId : String, aId : String) has the steps:

```
Lookup customer with customerId = cId;
Create a new account with accountId = aId;
Add this account to the customer accounts;
```

These are examples of normal behaviour, there may also be alternative sequences of actions for abnormal situations, such as where there is no account with *accountId* = *aId* in the *checkBalance*(*aId*) case.

Use cases can have *preconditions*, logical conditions that express in what situations the use case is valid to execute. E.g., *createAccount(cId : String, aId : String)* has the precondition that no account already exists with accountId = aId. Use cases can also have *postconditions*, defining their effect and result by a series of expressions.

4.3 OCL (Object Constraint Language)

OCL is the expression language used with UML, it can define logical conditions for preconditions, postconditions, invariants and other specification elements. OCL expressions have values of numeric, boolean, string, entity or collection types. The main use of OCL is to precisely define operation and use case functionalities.

The OCL standard numeric value types are *Integer* and *Real*. However in this book we will mainly use computational data types *int*, *long*, *double*. The string value type *String* ranges over sequences of characters. The usual numeric operators $+$, $-$, $*$, $/$, $<$, $>$, $\leq$, $\geq$ and functions *r*.*sqrt*(), *r*.*cos*(), *r*.*pow*(*p*), are present in the OCL library.

In UML-RSDS, mathematical functions of one argument can also be written without brackets: $r.sin.cos$, etc, when applied to simple expressions, or as $r \rightarrow f()$ for more complex expressions r, e.g., $(x + y * x) \rightarrow sqrt()$.

String functions include $s.size()$, $+$ (concatenation), $s.toLowerCase()$ and others. Boolean values are *true* and *false*, with the usual operators of conjunction &, disjunction *or*, negation *not*, and implication $\Rightarrow$. OCL specifies that short-circuit evaluation should be used for &, as in programming languages: if the first argument is false then the second argument is not evaluated. Similarly for *or*. A conditional expression *if* e *then* $e1$ *else* $e2$ *endif* returns the value of $e1$ if e is true, and the value of $e2$ if e is false.

Entity types can be used as follows in UML-RSDS:

- If E is a class diagram entity type, instances e of E can be used in OCL expressions, and feature values $e.att$, $e.role$ of e. Objects can be compared with $=$, $/=$
- A constraint with *context* E can refer directly to features of E. The context class is written before two colons, as:

  $$E::$$
  $$\quad\quad constraint$$

  Within *constraint* the features of E can be used without an object reference (implicitly they are features of the *self* object of E).
  For example:

```
Account::
     balance >= -overdraftLimit
```

  as an invariant of *Account*
- In the context of class E, the *self* object denotes an instance of E.
- The pre and postconditions of an instance-scope (non-static) operation of class E have context E.
- The inbuilt operation $createE() : E$ returns a new instance of a class E, for classes without an identity attribute.
- $createByPKE(v : String) : E$ looks up and returns (or creates and returns if no E instance with key v exists) the instance $E[v]$ of a class whose first identity attribute has value v.
- Objects of a class E can also be introduced by a *let* expression:

  $$let\ e : E = createE()\ in\ cond$$

Collection types include:

- $Set(T)$—the type of sets of T: a set $Set\{v_1, \ldots, v_n\}$ is an unordered collection of elements, with no duplicates (each element occurs only once).
- $Sequence(T)$—the type of sequences of T: a sequence $Sequence\{v_1, \ldots, v_n\}$ has elements in the listed order. Elements can occur multiple times.

E.g., $Set\{1, 9, 9, 1\}$ only has 2 elements: $Set\{1, 9\}$, whilst $Sequence\{1, 9, 9, 1\}$ has 4 elements.

Collection operators use the $\rightarrow$ symbol before the operator name:

- $s{\rightarrow}size()$ is the size of collection s
- $s{\rightarrow}sum()$ is the sum of elements of collection s (of numbers/strings)
- $s{\rightarrow}prd()$ is the product of elements of collection s (of numbers)
- $x : s$ is true if s contains element x, false otherwise. This is also written as $s{\rightarrow}includes(x)$.
- $s{\rightarrow}at(i)$ is the ith element of sequence s, which we also write as $s[i]$.

E.g., $9 : Set\{1, 9, 9, 1\}$ and $Sequence\{1, 9, 9, 1\}{\rightarrow}sum() = 20$.

Some collection operators are particularly useful in financial system specification:

- $s{\rightarrow}collect(x \mid e)$ is the sequence of values of expression e produced by applying e to the elements x in collection s. The order of the result will be the same of the order of s, if s is a sequence.

  This is akin to applying a function e to each member of a vector or matrix in Matlab, for example.
- $s{\rightarrow}select(x \mid P)$ is the subcollection of collection s which consists of all the $x : s$ that satisfy P.
- $s{\rightarrow}reject(x \mid P)$ is the subcollection of collection s which consists of all the $x : s$ that do not satisfy P.

The x argument in these operators can be omitted.

E.g., in the bank account example (Fig. 4.6),

$$accounts{\rightarrow}collect(balance)$$

is the sequence of *balance* values for a customer's accounts, and

$$accounts{\rightarrow}select(balance \geq 0)$$

are the accounts that are not overdrawn.

Operators can be chained, eg: $accounts{\rightarrow}collect(balance){\rightarrow}sum()$ is the sum of balances of accounts of a customer. Since $s{\rightarrow}collect(e)$ always produces a sequence of values, one for each element of s, duplicate values of e for different elements of s are distinguished and added correctly in such sums.

Quantifiers can also be applied to collections:

- $s{\rightarrow}forAll(x \mid P)$ is true if every element x of collection s satisfies P, false otherwise
- $s{\rightarrow}exists(x \mid P)$ is true if some element $x : s$ satisfies P, false otherwise
- $s{\rightarrow}exists1(x \mid P)$ is true if exactly one element $x : s$ satisfies P, false otherwise.

E.g., $Set\{1, 9, 9, 1\}{\rightarrow}forAll(x \mid x \leq 10)$ is true.

Sums and products over ranges of elements are defined by:

- *Integer.Sum*(a, b, i, e) corresponds to $\Sigma_{i=a}^{b} e$
- *Integer.Prd*(a, b, i, e) corresponds to $\Pi_{i=a}^{b} e$

The sequence of integers $a..b$ is written as *Integer.subrange*(a, b). This is *Sequence* $\{a, a+1, \ldots, b\}$.

In UML-RSDS, expressions can also be used to define behaviours:

- $x = v$ can be interpreted as "Set the value of x to v"
- $x : s$ can be interpreted as "Add x to collection s"
- *s*→*forAll*(x | P) as "Make P true for every element x of s"
- *E*→*exists*(x | P) for concrete entity type E as "Create an instance x of E and initialise it to satisfy P"
- *s*→*display*() displays the value of s
- *s*→*isDeleted*() removes the object or collection of objects s from the application.

This enables us to use OCL expressions to define the behaviour of operations and use cases, for example the *withdraw* operation definition given above. For use cases, postconditions of the use case can be written sequentially to define the steps of the use case in a logical manner, for example:

- *checkBalance(aId : String)*:

```
::
  aId : Account->collect(accountId)  =>
                Account[aId].balance->display()

::
  aId /: Account->collect(accountId)  =>
                ("No account with id = " + aId)->display()
```

This use case has two postcondition constraints, neither has an entity context, so their context is written as ::. The first postcondition executes in the case that there is an account with the given id value *aId* (normal behaviour of the use case), the second in the case that there is not (error behaviour).

- *createAccount(cId : String, aId : String)*:

```
::
  customer = Customer[cId]  =>
    Account->exists( a | a.accountId = aId &
                          a : customer.accounts )
```

This constraint looks up a *customer* : *Customer* Object by its id value and creates (or finds) an *a* : *Account* object with id = aId, and adds *a* to *customer*'s accounts. Instead of defining error behaviour for the case

$$cId \notin Customer \rightarrow collect(customerId)$$

we can assert a precondition for this use case:

```
::
   cId : Customer->collect(customerId)
```

It then becomes the responsibility of the caller of the use case to ensure that this precondition holds true at the point of call.

4.4 The Financial Specification Process

Given a new financial modelling or computational problem, the following steps can be followed (by an individual or team) to create a precise specification model for the problem:

- Carry out necessary background research to understand the context of the problem and the purpose of the task.
- Define an initial requirements specification based on financial concepts and relevant mathematics.
- Define (or reuse) a class diagram to represent the financial concepts and data involved in the problem.
- Describe informally the functional elements (operations and use cases) using the class diagram and mathematical theory as a basis.
- Formalise—express in a machine-readable form—the requirements specification and functional specifications to define exploratory prototypes operating on the data of the class diagram, which solve parts of the problem or simple cases of it.
- If the specification language is executable, then the prototypes can be executed and tested. Otherwise they can be checked by inspection and walkthroughs of their behaviour in particular scenarios.
- Progressively extend the prototypes to solve the complete problem, using real data where possible to test or check them.

At each stage, refer to stakeholder representatives, who should include (i) an expert in the specific finance domain; (ii) a customer representative who has detailed knowledge of how the application is intended to be used. One person may fulfil both roles (i) and (ii). Consideration should be given to stakeholders who may not be formally represented, but nonetheless are potentially affected by the application. It is particularly important to collaborate with a domain expert when defining the class diagram, which should accurately express domain concepts and terminology.

In the case of an executable specification language, the result of the above process is an executable specification which passes the user tests, but which may be inefficient, and may have poor structuring or other quality flaws. Further stages of refactoring (Sect. 4.7), redesign and optimisation may be needed before the application can be delivered.

4.5 Case Study: Estimating Internal Rate of Return (IRR)

Recall from Chap. 2 that the IRR or yield of a bond measures the quality of the bond as an investment: the effective rate at which the investment returns value over its term. We can model investments such as bonds by a class *Investment* which has an associated sequence of cash flows: for a bond these flows will consist of an initial payment (a negative cash flow), followed by coupon payments (positive cash flows to the investor), and repayment of the principal at end of the term (*futureValue* in Fig. 4.9). E.g.: an investor could invest £100 in a fixed-coupon bond for a term of 10 years, for a price (*presentValue*) of £110, receive 8% annual interest bi-annually (20 payments of £4), and £100 capital repayment at the end of the term.

We will record the coupon payments in the *flows* sequence, so that in terms of this model, the IRR is the (minimal) rate r such that:

$$presentValue = \Sigma_{i=1}^{flows.size} flows[i].amount/(1+r)^{flows[i].timePoint} + \\ futureValue/(1+r)^{flows.last.timePoint}$$

The right hand side is used as the definition of an operation

```
query totalValue(r : double) : double
pre: r /= -1
post:
  result = Integer.Sum(1,flows.size, i,
       flows[i].amount/((1 + r)->pow(flows[i].timePoint))) +
            futureValue/((1 + r)->pow(flows.last.timePoint))
```

which returns the total positive cash flows from the bond.

A version of the IRR equation using continuous compounding of interest is:

$$presentValue = \Sigma_{i=1}^{flows.size} flows[i].amount * e^{-r*flows[i].timePoint} + \\ futureValue * e^{-r*flows.last.timePoint}$$

Time can be measured in days, months, years, etc—r will be the rate with respect to this measure. Generally, r can be estimated by numerical approximation techniques, e.g., the bisection or secant methods, or by genetic algorithms. The above model and

Fig. 4.9 Investments class diagram

operation apply to any investment which has a series of cash flows, including fixed coupon and variable-coupon bonds.

We can give an informal specification of a use case to compute the IRR for a fixed coupon bond by listing the following steps:

1. Create an Investment instance *inv* with *presentValue* as the given price.
2. Create $term * frequency$ cash flows, each with the amount $coupon/frequency$ and time point $(i * 1.0)/frequency$ for the i'th flow. Add these in their time order to *flows*.
3. Set *futureValue* to the amount 100 representing the principal (the capital investment).
4. Apply the secant procedure from a suitable starting value $r0 > 0$ and $r0 < 1$, using the difference between *inv.presentValue* and *inv.totalValue*(r) as the function to minimise. This difference represents the net present value of the bond.

We can define the general *secant* procedure using the recursive operation:

```
static query secant(f : Function, rn : double,
  rminus : double, fminus : double, tol : double) : double
pre: tol > 0
post:
  fn = f.apply(rn) &
  (fn.abs < tol   =>  result = rn) &
  (fn.abs >= tol   =>
     result = secant(f, rn - fn*((rn-rminus)/(fn-fminus)),
                        rn, fn, tol) )
```

The parameter *rn* is the current best approximation to the root, *rminus* is the previous value of *rn* and *fminus* is $f(rminus)$.

This function is provided in the *NumericOptLib* UML-RSDS library (Fig. 4.10). The function approximates a root of f, ie., a value r such that $f(r) = 0$. Here, $f(rn)$ is $abs(inv.presentValue - inv.totalValue(rn))$ for an investment *inv*. *tol* is the tolerance or level of approximation to the root. Typical tolerance values are 0.001, 0.0001, etc. The smaller the tolerance, the more iterations of *secant* are needed.

The final step of the IRR use case can therefore be expressed as the following constraint, which computes initial values for the *secant* procedure using the *payout* of the bond (*totalValue*0), and invokes it using a *NetPresentValue* function:

```
Investment::
  totalValue0 = flows->collect(amount)->sum() +
                   futureValue &
  r1 = (totalValue0/presentValue)->pow(2.0/(flows.size + 1))
        - 1 &
  totalValue1 = totalValue(r1) &
  p = ((totalValue0 / presentValue)->log())  /
           ((totalValue0 / totalValue1)->log()) &
```

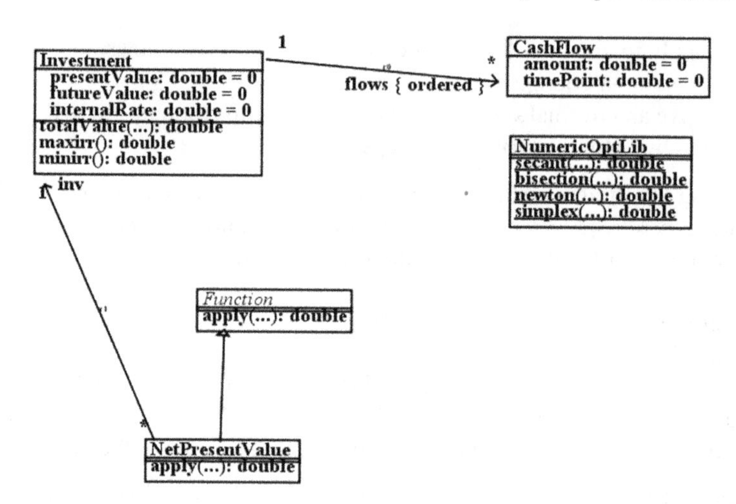

Fig. 4.10 Secant for IRR

```
r2 = (1 + r1)->pow(p) - 1  =>
   NetPresentValue->exists( f | f.inv = self &
      NumericOptLib.secant(f, r2, r1,
        totalValue1 - presentValue, 0.001)->display() )
```

NetPresentValue inherits from *Function* and has *apply* defined as:

```
query apply(x : double) : double
post:
  result = (inv.presentValue - inv.totalValue(x))->abs()
```

A specific instance f of *NetPresentValue* is created in the constraint, and *secant* is invoked with f as the first argument. Termination/convergence of secant is not guaranteed in general. Bisection or genetic algorithm methods can instead be used.

To test this application, a Java implementation of the UML-RSDS specification was used, and applied to a test case of 8 coupon bonds ranging from 1 year to 12 year terms:

```
BondId, Settlement, Maturity, Price, Coupon, Frequency
"1" , 1999 , 2000 , 103.78 ,   6.5 , 2
"2" , 1999 , 2001 , 106.72 ,   8.0 , 2
"3" , 1999 , 2002 , 112.58 ,  10.0 , 2
"4" , 1999 , 2003 ,  98.53 ,   5.5 , 2
"5" , 1999 , 2004 , 107.68 ,   8.0 , 2
"6" , 1999 , 2006 , 108.46 ,   8.0 , 2
"7" , 1999 , 2009 , 101.07 ,   7.0 , 2
"8" , 1999 , 2011 ,  93.11  , 6.0 , 2
```

The computed yield values for these are:

```
BondId, Yield
"1" , 0.0448200657634527
"2" , 0.06058085141979801
"3" , 0.07333535911340823
"4" , 0.05734245546660278
"5" , 0.06911033292597388
"6" , 0.07030851391741318
"7" , 0.06899785164137454
"8" , 0.06578688900172859
```

where 0.04 represents 4% yield, etc.

4.6 Case Study: Macaulay Duration of a Bond

The *Macaulay duration* of a bond is the time to maturity of the equivalent zero-coupon bond, or the weighted average time to payment. It weights the time of positive cash flows by their amount, to obtain a time point at which all the payments can be considered to occur together:

$$duration = (\Sigma_{i=1}^{flows.size} flows[i].timePoint * flows[i].amount * e^{-yield*flows[i].timePoint})/ (\Sigma_{i=1}^{flows.size} flows[i].amount * e^{-yield*flows[i].timePoint})$$

using continuous compounding, where we now include the principal repayment in *flows*.

This computation can use the previously-computed yield value (the IRR) of each bond, this is stored in *internalRate*. A version of the duration calculation using discrete compounding is:

$$duration = (\Sigma_{i=1}^{flows.size} flows[i].timePoint * flows[i].amount/(1 + yield)^{flows[i].timePoint})/ (\Sigma_{i=1}^{flows.size} flows[i].amount/(1 + yield)^{flows[i].timePoint})$$

The computation is formalised as an operation

```
query macaulayDurationC() : double
post:
  pv = totalValueC(internalRate) &
  dur = flows->collect( f | f.timePoint * f.amount *
            (-internalRate*f.timePoint)->exp() )->sum() &
  result = dur/pv
```

of *Investment*, for the continuous version, where *totalValueC* uses continuous compounding of the yield. A discrete compounding definition is:

```
query macaulayDuration() : double
post:
  pv = totalValue(internalRate) &
  dur = flows->collect( f | f.timePoint * f.amount /
            (1 + internalRate)->pow(f.timePoint) )->sum() &
  result = dur/pv
```

The computed durations for our example bonds are:

```
BondId, Yield, Duration
"1" , 0.0448200657634527  , 0.9844067709167799
"2" , 0.06058085141979801 , 1.8896640585964184
"3" , 0.07333535911340823 , 2.6770012951695747
"4" , 0.05734245546660278 , 3.6423011566865124
"5" , 0.06911033292597388 , 4.235571418580719
"6" , 0.07030851391741318 , 5.531443157037469
"7" , 0.06899785164137454 , 7.352752247451705
"8" , 0.06578688900172859 , 8.611712068377704
```

4.7 Specification Revision and Refactoring

Class diagrams and other models can be refactored to improve their structure, to remove redundancies and improve their correspondence to requirements. Refactoring is particularly important in an agile development approach, where it is usually applied on code. Some typical refactorings are [1]:

- "Pull up attribute" refactoring: if all (at least 2) direct subclasses of a class E declare an attribute $att : T$ with the same name and type, replace these declarations by a single definition of *att* in E (Fig. 4.11).
- "Move operation" refactoring: if an operation *op* of E refers to the attributes/roles of a linked class F via an association $E —_r F$, try moving *op* to F (Fig. 4.12) so that *op* can use these features without navigation via r.
- "Merge classes": if there are several direct subclasses of a class C, all empty, replace these subclasses by a flag attribute of C of enumerated type. If C has no further subclasses, make it a concrete class (Fig. 4.13). E.g., this transformation could be applied for the *Customer* class in Fig. 4.6.

'Move operation' is relevant to the IRR computation problem. In the definition of *totalValue* there are multiple references to features of *flows*[*i*] objects:

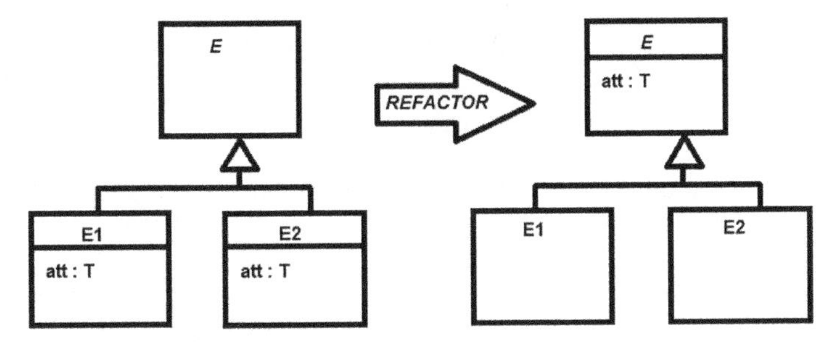

Fig. 4.11 Pull up attribute refactoring

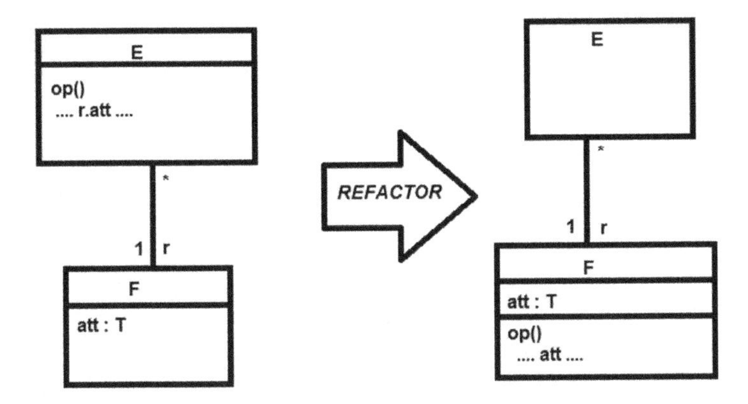

Fig. 4.12 Move operation refactoring

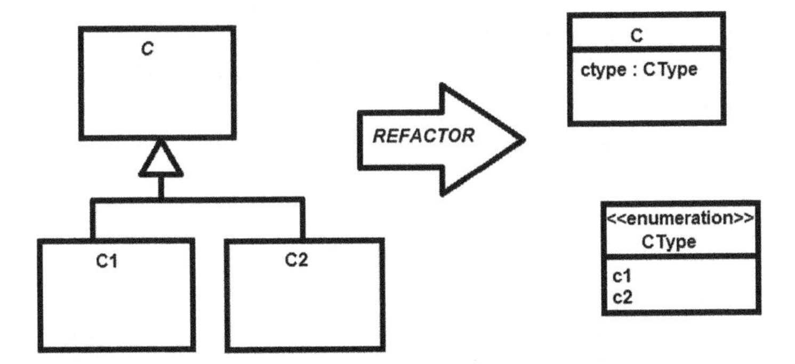

Fig. 4.13 Merge classes refactoring

```
query totalValue(r : double) : double
pre: r /= -1
post:
  result = Integer.Sum(1,flows.size, i,
   flows[i].amount/((1 + r)->pow(flows[i].timePoint))) +
           futureValue/((1 + r)->pow(flows.last.timePoint))
```

This can be improved by defining an operation *discountedAmount*(r : *double*) : *double* of *CashFlow* and moving the references to *flows*[i] features into this operation:

```
query discountedAmount(r : double) : double
pre: r /= -1
post:
  result = amount/((1 + r)->pow(timePoint))
```

totalValue can then be simplified to:

```
query totalValue(r : double) : double
pre: r /= -1
post:
  result = Integer.Sum(1,flows.size,i,
             flows[i].discountedAmount(r)) +
                futureValue/(1 + r)->pow(flows.last.timePoint)
```

Summary

In this chapter, we have:

- Introduced the essential UML class diagram notations.
- Introduced use case concepts.
- Introduced core OCL features and uses.
- Illustrated financial system specification with two small examples based on real-world problems.
- Considered class diagram model refactoring.

Exercises

1. How would the banking example be modified if accounts could only be associated with 3 customers at most?

2. Use a $\rightarrow$*select* or $\rightarrow$*reject* to express the collection of a customer's accounts which are overdrawn, ie., with $balance < -overdraftLimit$.

3. Extend the class diagram of Fig. 4.6 to include a class *Transaction*: for each account there is a sequence of transactions, which record the kind of operation (*withdraw* or *deposit*) and the amount of the transaction. Each transaction can be for 1 or 2 accounts, a *main* account is the account on which the action takes place, a *secondary* account is present if funds are being transferred from/to another account in the same bank.

4. Give the definition in OCL of an operation $sumsqdiffs(s1 : Sequence(double), s2 : Sequence(double)) : double$ which computes the sum of squared differences $\Sigma_{i=1}^{s1.size}(s1[i] - s2[i])^2$ of two sequences of doubles, of the same length.

5. Instead of using secant, another technique for finding the IRR is bisection: dividing a range of possible values in half repeatedly until a solution is found (within a degree of precision):

```
static query bisection(f : Function, r: double,
                       rl: double, ru: double,
                       tol : double): double
pre: true
post: v = f.apply(r) &
  ( ( ru - rl < tol => result = r ) &
    ( ru - rl >= tol & v > 0 =>
          result = bisection(f,( ru + r ) / 2,r,ru,tol) ) &
    ( ru - rl >= tol & v < 0 =>
          result = bisection(f,( r + rl ) / 2,rl,r,tol) ) &
    ( true => result = r ) )
```

f is assumed to be decreasing on the interval $[rl, ru]$. Does this approach always converge for the function f defined as the net present value of an investment?

Assuming that $ru = 1$ and $rl = 0$ initially, how many iterations are needed to reach a result in the case that there is convergence?

6. How do you think the use of executable specifications helps to ensure the correctness of an application, compared to non-executable specifications?

7. What is the advantage of refactoring at a (platform-independent) specification level, compared to refactoring at the code level?

Reference

1. M. Fowler, K. Beck, J. Brant, W. Opdyke, D. Roberts, *Refactoring: Improving the Design of Existing Code* (Addison-Wesley, USA, 1999)

Chapter 5
Financial System Design

This chapter will describe the design of financial software using an agile MBD process. We will consider software design quality, design patterns and software reuse, and describe the QuantLib finance library.

5.1 Agile Development Process

In general, we recommend the use of an agile model-based development process for financial system development, based on the Scrum process or other agile approach, but with work focussed on writing specifications, not code (Fig. 5.1). Product and sprint backlogs are used, and for each work item within a sprint there are three stages of development: (i) requirements and informal specification; (ii) formal specification and design; (iii) integration and testing.

In agile approaches, the customer representative has a key role in ensuring that what is being developed actually meets the customer and user needs. Ideally, they participate directly in the development, giving immediate feedback on models and prototypes.

Scenario analysis and executable prototyping can be used at stages (i) and (ii) to identify the specific functionalities and behaviours required from the system, in conjunction with the customer representative. Formal specification and design consists of writing a precise machine-readable version of the informal specification, and organising this in a modular manner. For example, defining small cohesive units of functionality as operations within particular classes, and allocating responsibilities to classes. Components consisting of several classes and providing operations through a public interface can also be defined. Such components should ideally be suitable for reuse in other contexts and applications. For example, a component for computing properties of investments, such as yields.

K. Lano and H. Haughton, *Financial Software Engineering*, Undergraduate Topics in Computer Science, https://doi.org/10.1007/978-3-030-14050-2_5

The specification can be validated by inspection, walkthroughs and testing, and refactored to improve its structure, as discussed in the previous chapter.

If a new requirement is introduced which affects existing functionalities, then impact analysis for this change is also carried out in stage (i). At each stage we look for reuse opportunities from existing libraries or previous applications in a product family, and at stages (ii) and (iii) we can consider contributions of operations or components from this development to the library. In addition to the usual Scrum roles of development team, customer representative and Scrum leader, additional roles of a MBD expert and a specification library manager are needed. For projects requiring supporting tool development, an additional tooling team is needed to work in liaison with the main project team to provide support tools and enhancements needed by the team. For example, tools to provide necessary data format conversions or specialised code generation capabilities. The tool support team also works in an agile MBD manner to produce the necessary tools (right hand side of Fig. 5.1). Their customer representative is from the main development team (left hand side of Fig. 5.1).

Generally, the scope of a finance project will include a number of financial domains, such as option pricing, bond pricing, etc. There will usually be existing mathematical models of the domains which may need to be adapted or specialised for the project. Class diagram models and libraries of operations for the domains may already exist prior to a project, in which case they can be reused and extended as necessary. Otherwise, a new model of the financial and computational elements will need to be created. In the case of work items which are financial procedures, stage

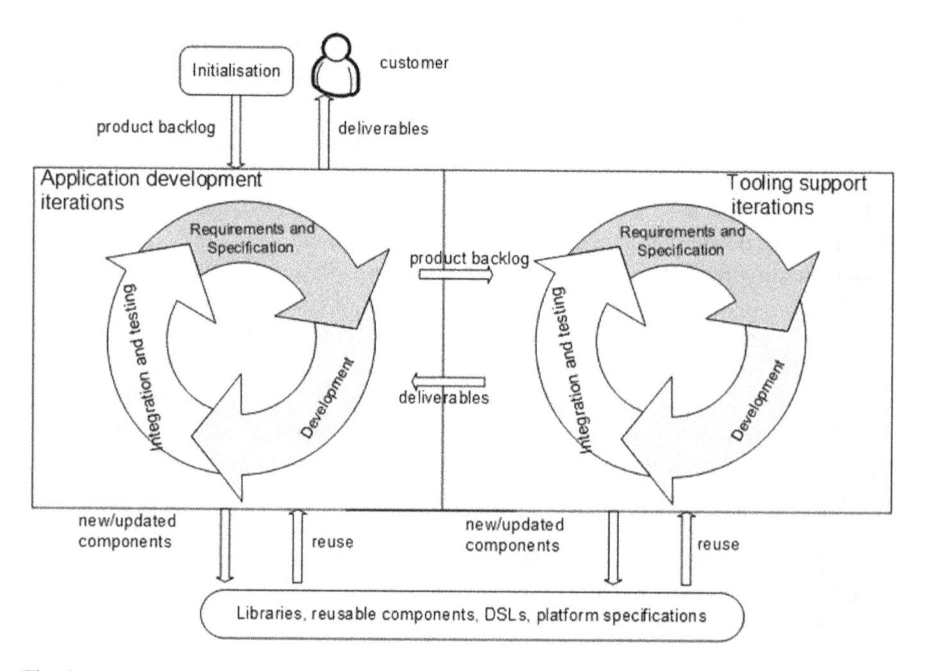

Fig. 5.1 UML-RSDS agile MBD process

(i) involves eliciting functional and non-functional requirements from all stakeholders, and expressing the steps of the procedure in natural language, with terminology based upon the class diagram and mathematical models. In stage (ii) this specification is reviewed and formalised as a UML-RSDS use case, where the steps become successive postcondition constraints written in OCL. These constraints typically use operations from the classes of the class diagram.

In stage (iii) code is generated from the specification and tested. The specification is then revised as necessary until the requirements are met, including non-functional requirements such as efficiency and modularity.

5.2 Optimisation

We recommend the use of the specification of a system to also define its design and implementation. This has many advantages:

- The implementation can be tested at the same time that the specification is validated—the two processes are the same.
- A single artifact is used to define the system, instead of separate specifications and code, which would need to be maintained together and kept in alignment.
- A system implementation can be produced for any platform which has an available code generator. It may be necessary to write a new generator or to adapt an existing generator.
- The system can be maintained and enhanced by changing the specification, a task that is usually much faster than changing code.
- Refactorings and optimisations applied to the specification will be automatically applied to any implementation.

However, a disadvantage is that the clarity of the specification is sometimes in conflict with its efficiency, hindering the use of the same artifact for both specification and code generation. In UML-RSDS the architecture of generated code is fixed, and cannot be modified by the developer. However they can organise the definitions of use cases, classes and operations as they wish.

In general, to optimise a specification, we recommend:

- Avoid the use of recursive operation definitions where possible. Provide a non-recursive *activity* definition in cases where the specification (operation postcondition) is recursively defined.
- If query operations are frequently called with the same arguments, make the operation *cached*: this means that computed results are stored in a map and looked-up instead of being recomputed on subsequent calls. This applies for operations with discrete-valued arguments, and may be less beneficial for operations with *double*-valued arguments.

Using these techniques, we have found that it is possible to produce generated code that is more efficient than hand-crafted code for the same application [1].

5.3 Case Study: Bootstrapping of Interest Rates

Interest rate *bootstrapping* is the process of calculating interest rates for all periods over a given range, starting from a number of known rates. Bootstrapping is one technique for creating a yield curve from market data, however it can be less systematic than approaches which fit a curve based on a parameterised formula (such as the NS or NSS models) to the data.

In general, a coupon bond with present value *price* and term m years will make $m * frequency$ regular coupon payments (positive cash flows) over its term, together with repayment of the principal at the end of the term. We can generalise this situation to consider any *Investment* with a present value (price) and future value (the principal repayment at term), and a sequence of cash flows of an *amount* at a *timePoint* (Fig. 5.2).

Each payment is discounted by a factor $flows[i].discount$ dependent upon the annual interest rate $flows[i].rate$ for the period 0 to $flows[i].timePoint$:

$$\begin{array}{l} CashFlow :: \\ \quad discount = 1/(1 + rate) \rightarrow pow(timePoint) \end{array}$$

and therefore

$$\begin{array}{l} CashFlow :: \\ \quad rate = discount \rightarrow pow(-1/timePoint) - 1 \end{array}$$

Because the present value (price) should be equal to the sum of discounted payments, we have the equation:

$$\begin{array}{l} presentValue \; = \\ \qquad Integer.Sum(1, n-1, i, flows[i].amount * flows[i].discount) + \\ \qquad (100 + flows[n].amount) * flows[n].discount \end{array}$$

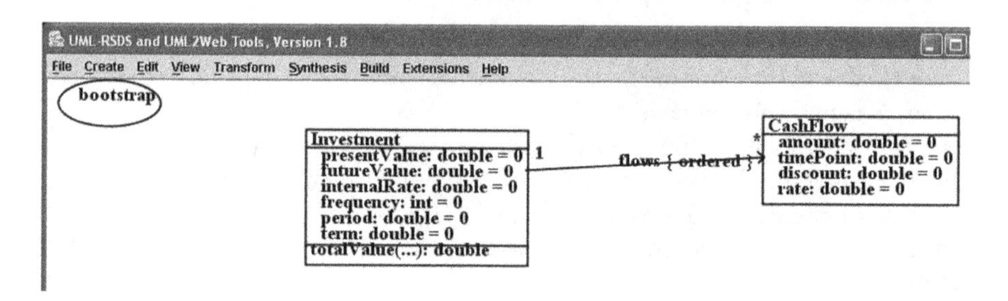

Fig. 5.2 Extended investments class diagram

where $n = flows.size$. Using this equation we can deduce from the rates and discount factors for $i = 1 \ldots n-1$ the discount factor for n, which can be computed as:

$$\begin{aligned} flows[n].discount \; = \; & \\ & (presentValue - Integer.Sum(1, n-1, i, \\ & \quad flows[i].amount * flows[i].discount))/ \\ & \qquad (100 + flows[n].amount) \end{aligned}$$

Recall that $Integer.Sum(a, b, i, e)$ computes the sum $\Sigma_{i=a}^{b} e$ where e depends on i. The rate for n can therefore be derived from known rates for 1 up to $n-1$, and used to derive further rates when bootstrapping is applied to investments with longer terms.

Therefore, if the discounts for the time points $flows[1].timePoint$, ... $flows[n-1].timePoint$ are already known, for bonds with the same settlement (starting) date and provided by the same issuer, we can define a financial procedure for bootstrapping the discount and rate for $flows[n].timePoint$, using a fixed coupon bond from the same issuer and with the same settlement date, as follows:

1. Create an *Investment* instance and initialise its *presentValue*, *term*, *frequency* and *period*.
2. Create the positive cash flows of the bond from the coupon payments for the term. The amount of each coupon payment is $coupon/frequency$ and the time of the ith payment is $i * 1.0/frequency$.
3. Initialise the known discounts and rates of the flows from $1 \ldots n-1$.
4. For the flow at n compute its discount and rate using the bootstrapping formula.

Assumptions made are that the repayment made at the end of each term is 100 (i.e., 100% of the principal) and that the price, frequency and known rates are all positive. We assume also that the bond is newly issued and does not have accrued interest.

The formalised specification is then defined by the postconditions of a use case *bootstrap*.

The input parameters of the use case are *price*, *term*, *coupon*, all double values, and integer *frequency*. An input parameter $known : Sequence(double)$ with $known.size = floor(term * frequency) - 1$ holds the known interest rates. The use case has postconditions:

```
  ::
   Investment->exists( b | b.presentValue = price & b.term = term &
                           b.frequency = frequency & b.period = 1.0/frequency )

 Investment::
   i : Integer.subrange(1,  (term*frequency)->floor())   =>
         CashFlow->exists( f | f.amount = coupon/frequency &
                                 f.timePoint = i*period & f : flows )

 Investment::
   i : Integer.subrange(1, known.size)   =>
```

```
        flows[i].rate = known[i] &
        flows[i].discount = 1/((1 + known[i])->pow(flows[i].timePoint))

Investment::
  n = flows.size  =>
        flows[n].discount =
            (presentValue - Integer.Sum(1, n-1, i,
                        flows[i].amount * flows[i].discount@pre) ) /
                                    (100 + flows[n].amount )

Investment::
  n = flows.size  =>
         flows[n].rate = flows[n].discount->pow(-1/flows[n].timePoint) - 1
```

The use case constraints are similar to the finance theory equations, and simply express these in OCL notation. Thus it is direct to validate the specification against the theory. Note that *discount*@pre is used in the fourth constraint because *discount* is both read and written in the constraint, so the *pre* annotation is needed to avoid a fixed-point implementation being used. We can avoid this problem by factoring the inner computation into an operation of *CashFlow*, as with the IRR case study.

The outcome of the bootstrapping process is that interest rates can be inferred for all annual durations up to the longest bond term in the market data set. For durations which fall between the annual time points of market data items, linear interpolation of the yields can be used. A similar bootstrapping procedure can be used for other financial products.

5.4 Libraries and Reuse

During development, a project team may identify useful functionalities and components which could be reused in future projects. Such functions and components can be added to appropriate library classes and models. There are two different ways of defining libraries with MBD: (i) platform-independent explicit specifications of the library components can be given, from which implementations in different programming languages can be automatically generated; (ii) the libraries cannot be specified in a platform-independent manner, e.g., because they use platform-specific aspects such as graphics or file system operations. In this case the component operations can be given an interface specification (only their signatures are defined), and separate implementations are provided for each platform (e.g., by hand-written code). Of course, since this involves more manual coding effort, option (i) is preferable where possible.

An example of the first case is the identification of mathematical functions $isPrime(n : int)$, $factorial(n : int) : long$, $combinatorial(n : int, m : int) : long$ "n choose m", $gcd(n : int, m : int) : int$ and $lcm(n : int, m : int) : int$ as useful for potential reuse during a project. These operations could then be defined in a language-independent *mathlibmm.txt* model to include in future applications.

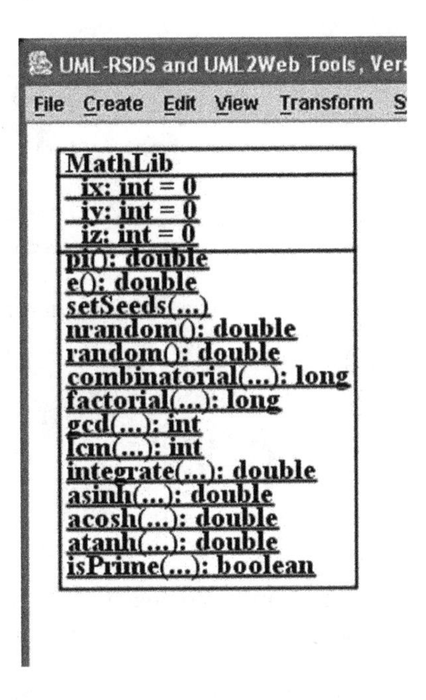

Fig. 5.3 Mathematical functions library

A class *MathLib* can be created with these functions as static operations (Fig. 5.3). Since the functions are to be widely used, care should be taken to ensure their efficiency and correctness.

isPrime is defined as:

```
static query isPrime(n: int): boolean
pre: true
post:
  (n < 2 => result = false) &
  (n = 2 => result = true) &
  (n > 2 =>
      result = Integer.subrange(2, n.sqrt.floor)->forAll(
                                              i | n mod i > 0 ))
```

In the final clause, n is considered prime if no number between 2 and *n.sqrt.floor* divides it.

For the *factorial* and *combinatorial* operations, the *Integer.Prd*(a, b, i, e) operator can be used to compute the product $\Pi_{i=a}^{b} e$, to optimise these operations, instead of using recursion. The precondition $x \leq 20$ is needed for factorial to prevent numeric overflow, since the factorial of 21 is larger than the maximum long value. Likewise for combinatorials.

```
static query combinatorial(n: int, m: int): long
pre: n >= m & m >= 0 & n <= 25
post:
  (n - m < m =>
```

```
    result = Integer.Prd(m+1,n,i,i)/Integer.Prd(1,n-m,j,j)) &
   (n - m >= m =>
    result = Integer.Prd(n-m+1,n,i,i)/Integer.Prd(1,m,j,j))

static query factorial(x: int): long
pre: x <= 20
post:
    ( x < 2 => result = 1 ) &
    ( x >= 2 => result = Integer.Prd(2,x,i,i) )
```

For gcd, there is a well-known recursive computation:

```
static query gcd(x: int, y: int): int
pre: x >= 0 & y >= 0
post:
    (x = 0   =>   result = y) &
    (y = 0   =>   result = x) &
    (x = y   =>   result = x) &
    (x < y   =>   result = MathLib.gcd(x, y mod x))  &
    (y < x   =>   result = MathLib.gcd(x mod y, y))
```

Such recursive computations should be avoided where possible in libraries, for efficiency and robustness reasons, and instead replaced by explicit algorithms:

```
static query gcd(x: int, y: int): int
pre: x >= 0 & y >= 0
post: true
activity:
    l : int ; k : int ; l := x  ; k := y  ;
    while l /= 0 & k /= 0 & l /= k
    do
        if l < k then k := k mod l
        else l := l mod k ;
    if l = 0 then result := k
    else result := l ;
    return result
```

The recursive definition can be retained as documentation, but the activity will be used for code generation. *combinatorial*, *factorial* and *gcd* can be defined as *cached* operations, for additional efficiency. Note that an explicit declaration of *result* is not needed in the activity of a query operation.

From the gcd, the lcm can be directly calculated:

```
static query lcm(x: int, y: int): int
pre: x >= 1 & y >= 1
post: result = ( x * y ) / MathLib.gcd(x,y)
```

We can also improve the efficiency of *isPrime* by defining an activity for it:

```
static query isPrime(n: int): boolean
pre: true
post:
  (n < 2 => result = false) &
  (n = 2 => result = true) &
```

```
  (n > 2 =>
    result =
        Integer.subrange(2, n.sqrt.floor)->forAll( i |
                      n mod i > 0 ))

activity:
  if n < 2 then return false
  else if n = 2 then return true else skip ;

  b : int := n.sqrt.floor ;
  i : int := 2 ;
  while i <= b
  do
    if n mod i = 0 then return false
    else i := i+1 ;
  return true
```

An example of the second kind of library is the XMLParser component, which enables a UML-RSDS application to extract data from XML format files. A Java implementation is provided for this component.

Some existing UML-RSDS libraries include:

- *Real*, with operations *subrange*(*low* : *double*, *step* : *double*, *upper* : *double*) : *Sequence*(*double*), *minValue*() : *double*, *maxValue*() : *double*
- *MathLib*, with *random*, *factorial*, *combinatorial*, *gcd*, *lcm*, *pi*, etc., as above.
- *NormalDist*, with operations *normal*, *cumulative*, *sample*.
- *Matrix*, with operations on (2-dimensional) matricies (Fig. 5.4).

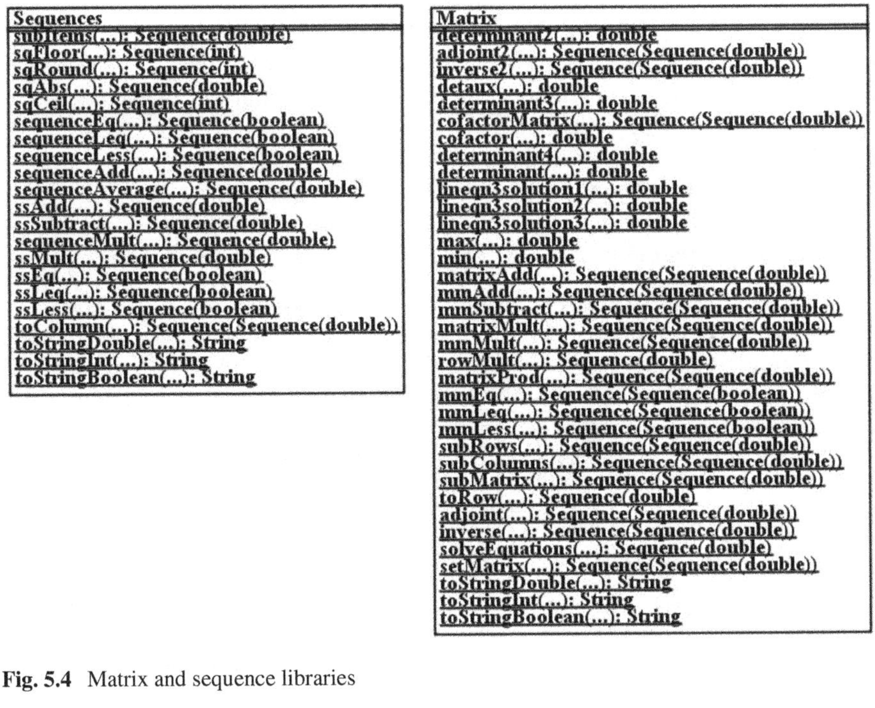

Fig. 5.4 Matrix and sequence libraries

- *Sequences*, with operations on sequences (Fig. 5.4).
- *StatLib*, with operations $mean(Sequence(double)) : double$, *mode*, etc.
- *NumericOptLib*, with operations for *secant*, *bisection* and other numerical optimisations.
- *StringLib*, with operations $before(str : String, delim : String) : String$, $after(str : String, delim : String) : String$, $split(str : String, delim : String) : Sequence(String)$, etc.
- XMLParser, to parse XML documents.

These libraries are defined in *realmm.txt*, *mathlibmm.txt*, etc, at https://nms.kcl.ac.uk/kevin.lano/libraries and in the umlrsds distribution.

5.5 Design Quality Flaws

Object-oriented code and models can have quality flaws, sometimes referred to as 'bad smells', which are poor-quality structures and organisation of the system that can impair understanding and make the system difficult to change. Although these are not functional errors, such flaws can lead to higher costs, especially in an agile approach where the ability to change models or code is essential.

Examples of quality flaws include:

- *God Class*: one class carries out most of the system functionality, whilst others are auxiliary to it.
- *Excessive Class Length*; *Excessive Method Length*; *Excessive Inheritance use*.
- *Excessive Parameter List* (e.g., more than 10 parameters in an operation).
- *Duplicate Code*: sections of identical code in different locations.
- *Cyclomatic Complexity* (a high number of logical conditions in an operation, representing different possible execution paths).
- *Too Many Methods/Operations* (e.g., more than 20 in a class).
- *Too Many Fields/Attributes* (e.g., more than 20 in a class).
- *Excessive Fan-Out* (more than 5 calls to different operations from one operation), *Excessive Fan-In* (more than 5 different callers of one operation).

Code/design smells can make maintenance of the system more expensive, and increase the likelihood of functional errors being introduced or remaining undetected. For example, an excessively large class or operation is difficult to understand or change, whilst an operation with high cyclomatic complexity is difficult to test. Duplicate Code means that any change in one copy must also be duplicated to the others. Refactoring can be used to remove these flaws and improve structure.

5.5.1 Technical Debt

Technical debt (TD) refers to the short and long-term impact of software quality flaws [2]. The *principal* cost of TD is incurred when refactoring or other redesign is used to remove the flaw from the software (analogously to paying off a debt by paying back the principal of the loan). Alternatively, *interest* on the debt is paid in additional costs due to the flaw, each time the software is maintained.

Technical debt is a problem both for manually-written code and automatically-generated code [3]. We consider that this issue should be addressed during development as part of the agile practice of regular refactoring. If TD is addressed and reduced at the specification level, this should also result in reduced TD in generated code, provided that the code generators are constructed to avoid introducing additional TD such as code duplication.

5.5.2 Measuring Technical Debt

Tools such as PMD (https://pmd.github.io) can be used to identify design flaws in code. PMD identifies class and method size issues, cyclomatic complexity, etc. PMD uses the following design flaw thresholds for code, based on lines-of-code (LOC) (Table 5.1).

Similar indicators are used in the SonarQube tool for the SQALE method [4], which counts occurrences of code duplication and excessive method complexity. Code duplication is an indicator of inadequate factoring and generalisation of functionality within the code. A CC limit of 10 is also suggested for Java in the SQALE method examples of [5], although with a lower EPL threshold of 5.

Such thresholds can also be adopted as indicators of quality flaws in UML specifications. Instead of LOC as a measure of class and operation size, a measure c of semantic complexity of expressions and activity pseudocode can be used. Tables 5.2 and 5.3 show how this is defined, together with a count of number of tokens, t.

The UML-RSDS tools use c as a basis for identifying quality flaws and estimating technical debt, with a threshold of $c = 100$ for individual operations, and $c = 1000$ for a complete specification. Duplicate expressions or activity code with $t > 10$ tokens are also identified.

Table 5.1 Typical thresholds for code flaws

Code smell	Threshold
Excessive class length (ECS)	1000 LOC
Excessive method/operation length (EOS)	100 LOC
Excessive parameter list (EPL)	10 parameters
Cyclomatic complexity (CC)	10
Too many operations/methods (ENO)	10 per class
Too many attributes/fields (ENA)	15 fields

Table 5.2 OCL expression complexity measures

Expression e	Complexity $c(e)$	Token count $t(e)$
Numeric, boolean or String value	0	1
Identifier *iden*	1	1
Basic expression $obj.f$	$c(obj) + c(f) + 1$	$t(obj) + t(f) + 1$
Operation call $e(p1, \dots, pn)$	$c(e) + 1 + \Sigma_i c(pi)$	$t(e) + n + 1 + \Sigma_i t(pi)$
Unary expression $op\ e$ $e \rightarrow op()$	$1 + c(e)$	$1 + t(e)$ $4 + t(e)$
Binary expression $e1\ op\ e2$ $e1 \rightarrow op(e2)$	$c(e1) + c(e2) + 1$	$t(e1) + t(e2) + 1$ $t(e1) + t(e2) + 4$
Ternary expression $op(e1, e2, e3)$ $if\ e1\ then\ e2$ $else\ e3\ endif$	$c(e1) + c(e2) + c(e3) + 1$	$t(e1) + t(e2) + t(e3) + 5$ $t(e1) + t(e2) + t(e3) + 4$
$let\ v : T = e1\ in\ e2$	$c(T) + c(e1) + c(e2) + 4$	$t(T) + t(e1) + t(e2) + 5$
$Set\{e1, \dots, en\}$ $Sequence\{e1, \dots, en\}$	$1 + \Sigma_i c(ei)$	$2 + n + \Sigma_i t(ei)$

Table 5.3 Activity complexity measures

Activity s	Complexity $c(s)$	Token count
`return` e	$1 + c(e)$	$1 + t(e)$
$v := e$	$c(v) + c(e) + 1$	$t(v) + t(e) + 1$
$s1;\ s2$	$c(s1) + c(s2) + 1$	$t(s1) + t(s2) + 1$
Operation call $e(p1, \dots, pn)$	$c(e) + 1 + \Sigma_i c(pi)$	$t(e) + n + 1 + \Sigma_i t(pi)$
`if` e `then` $s1$ `else` $s2$	$1 + c(e) + c(s1) + c(s2)$	$3 + t(e) + t(s1) + t(s2)$
`for` $v : e$ `do` s	$c(e) + c(s) + 1$	$3 + t(e) + t(v) + t(s)$
`while` e `do` s	$c(e) + c(s) + 1$	$t(e) + t(s) + 2$
`break`	1	1
`continue`	1	1
`var` $v : T$	$c(T) + 3$	$t(T) + 3$

5.5.3 *Refactoring of Code and Models*

Refactoring is the process of modifying code or models to improve structure, whilst keeping the same semantics. For example, in order to remove flaws and reduce technical debt:

- Large methods/operations can be factored into smaller parts.
- Duplicated code can be factored into a new operation, which is called from the locations of the copies.
- Multiple parameters for a operation can be bundled into a single object parameter.

A schematic example of factoring out duplicated activity code could be as follows. The original code could be:

```
m1()
activity:
  ... m1-specific code ...
  ... duplicated code s1; s2; s3; ...


m2()
activity:
  ... m2-specific code ...
  ... duplicated code s1; s2; s3; ...
```

The refactored code could be:

```
mnew()
activity:
  ... duplicated code s1; s2; s3; ...

m1()
activity:
  ... m1-specific code ...
  mnew()

m2()
activity:
  ... m2-specific code ...
  mnew()
```

This refactoring reduces the size of the operations, and removes duplicated code. However it increases the number of operations and may increase the number of calls to operations.

Parameter bundling forms a single object to represent a group of related parameters, thus removing an EPL flaw.

The original activity code could have the form:

```
m(x : X, y : Y, p1 : T1,..., pn : Tn)
activity:
  ... m's code ...
```

Assume that $p1 \ldots pn$ are related data, e.g., different attributes of a *Person*.

Define (or use) a class *PClass* that groups these items together:

```
class PClass
{ attribute p1 : T1;
  ...
  attribute pn : Tn;
}
```

The refactored code is then:

```
m(x : X, y : Y, p : PClass)
activity:
  ... m's code, access to pi is now p.pi ...
```

This reduces the number of parameters, and groups related data into a class.

5.6 Design Patterns

Design patterns are structures of software which define solutions to particular design problems. Patterns are mainly independent of specific programming languages, although they are aligned more to object-oriented languages. They can also be used for UML.

There are three main categories of design patterns:

Creational: these organise the creation of objects and of object structures. E.g.: *Singleton*, *Factory*

Behavioural: these organise the execution of behaviour amongst objects. E.g.: *Iterator*, *Observer*, *Strategy*

Structural: these organise the structure of classes and relationships. E.g.: *Proxy*, *Facade*.

5.6.1 Singleton Pattern

This is a creational pattern used to define classes which should have only a single instance (for example, a single point of access to a resource such as shared repository).

Singleton is used when there must be a unique instance of a given class, accessible to clients from a well-known access point.

The involved classes are:

- *Singleton*—defines an operation *instance* that lets clients access its unique instance. *instance* is a class-scope (static) operation.
- *Singleton* may also be responsible for creating its own unique instance.

A typical schematic structure of a singleton class, in Java, is:

```
public class Resource
{ private static Resource uniqueInstance = null;
  ...

  private Resource() { }

  public static Resource getResource()
  { if (uniqueInstance == null)
    { uniqueInstance = new Resource(); }
    return uniqueInstance;
```

```
    }

    public boolean request(Data d)
    { ... }
}
```

Because the constructor is private, only the *Resource* class itself can construct *Resource* instances: this is only done once, when *getResource* is called for the first time.

A client uses the singleton via calls

```
Resource.getResource().request(x);
```

5.6.2 Observer Pattern

Observer is a behavioural pattern for the management of multiple dependent objects whose state may need to change as a result of changes to a resource that they depend upon. The dependent objects are termed *observers* or *views* of the resource, which is termed the *observable* or *subject*.

Observer is a means by which logical invariants which span different objects can be maintained (Fig. 5.5).

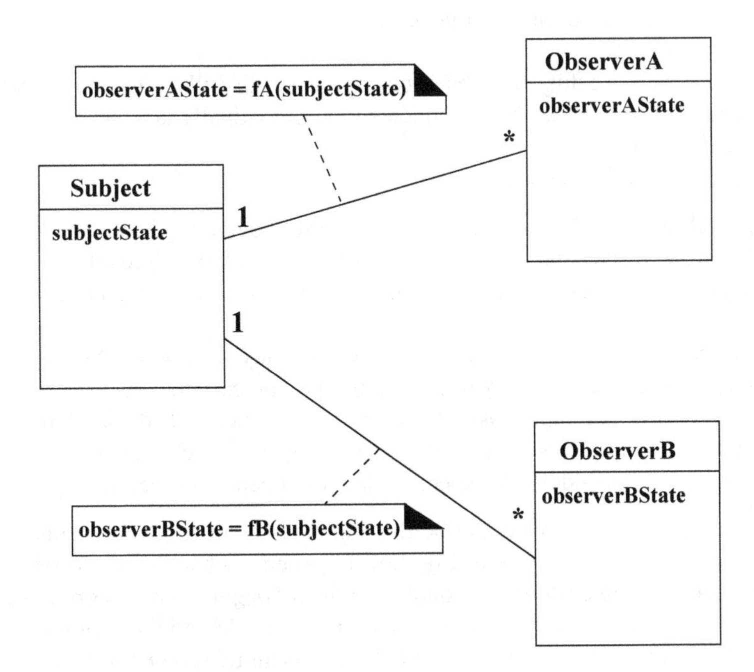

Fig. 5.5 Abstract observer pattern

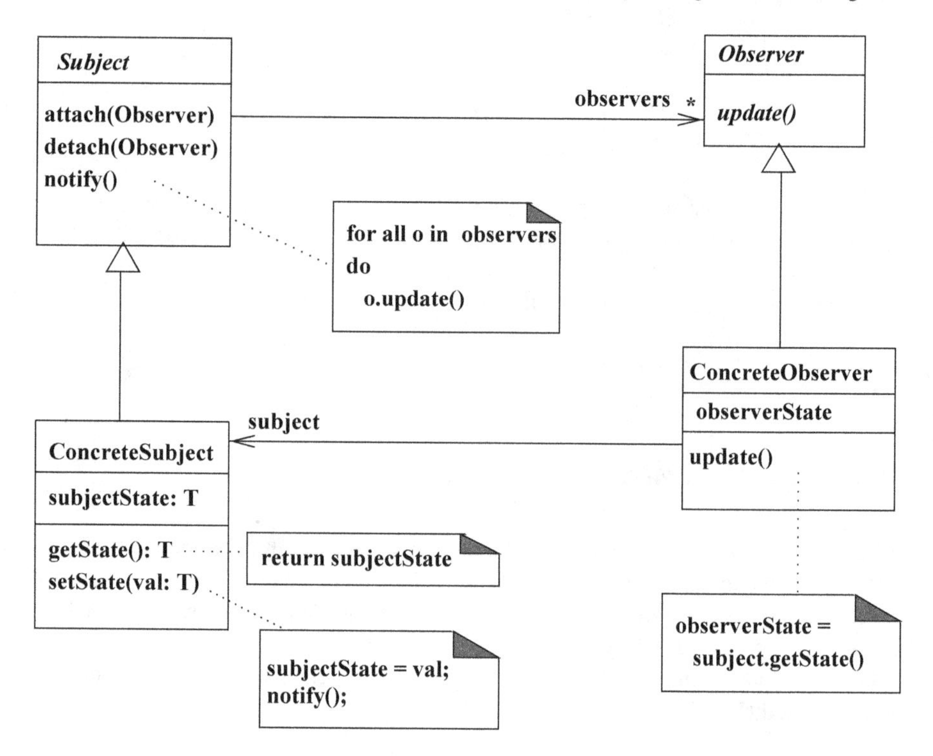

Fig. 5.6 Design structure of observer pattern

In order to achieve this, the subject object must notify all its dependents when its data is changed (Fig. 5.6). The dependents then individually take necessary actions to update their own state.

The participants in the pattern are:

- *Subject* class—the abstract superclass of classes containing observed data. It has methods *attach* and *detach* to add/remove observers, and *notify* to inform observers that a state change occurred on the observable, so they may need to update their data.
- *ConcreteSubject*—defines specific observables, any method of this class which modifies the subject data may need to call *notify* on completion.
- *Observer*—abstract superclass of observers of subjects. It declares the *update* method to adjust the observer's data on any subject state change.
- *ConcreteObserver*—defines a specific class of dependent objects.

An example of Observer could be the maintenance of logs of account transactions in the bank account system (Fig. 4.6). In general, a bank needs to retain a log of all transactions on customer accounts. Define a *Logger* class to store a sequence of transaction records, each transaction has an *accountId* and the action performed (*withdraw*, *deposit*, etc) on that account. *Logger* is an *Observer* for *Account* (playing the role of a *Subject*). The account *notify* operation is invoked by *deposit* and

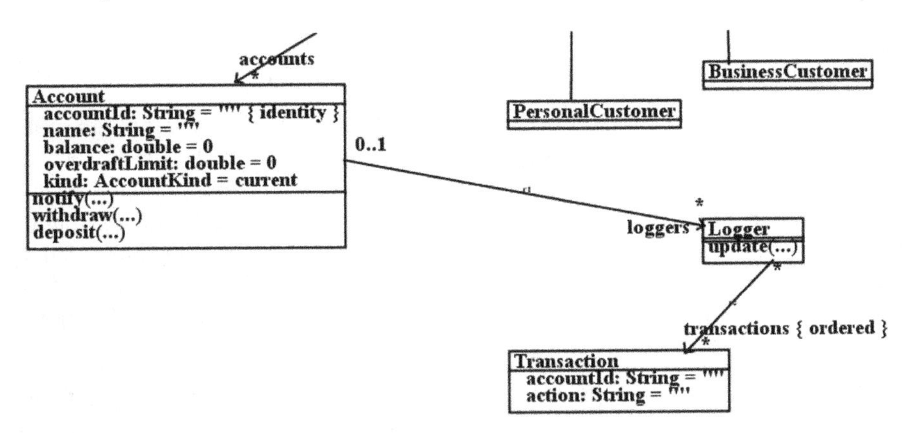

Fig. 5.7 Observer pattern for account logging

withdraw, and sends transaction data to all attached *Logger* objects, via *update* invocations on these. Each logger responds to *update*(*aId*, *data*) by creating a new *Transaction* and storing this in its log sequence (Fig. 5.7).

The *Account* updater operations call *notify*:

```
Account::
deposit(amt : double)
pre: amt >= 0
post:
  balance = balance@pre + amt &
  notify("deposit " + amt)

Account::
withdraw(amt : double)
pre: balance - amt  >=  -overdraftLimit
post:
  balance = balance@pre - amt &
  notify("withdraw " + amt)
```

notify sends the updated data to loggers:

```
Account::
notify(s : String)
post:
  loggers->forAll( lg | lg.update(accountId, s) )
```

These then create a new transaction and add it to their log sequence:

```
Logger::
update(id : String, s : String)
post:
  Transaction->exists( t | t.accountId = id &
                          t.action = s & t : transactions )
```

Several different loggers could be attached to each account, e.g., for customer services purposes as well as for regulatory record-keeping or fraud detection purposes.

5.6.3 Iterator Pattern

This behavioural pattern addresses the common situation of iteration through some collection of elements, processing each in turn. Collections can be linear, such as sequences and arrays, acyclic such as trees, or more general graphs or networks.

The pattern separates the iteration process into a *Iterator* object, which at a minimum supports operations to initiate the iteration (e.g., *start*()), to step one position forward (e.g., *next*()), to obtain the element at the current position (e.g., *element*()), and to test if the end of the iteration has been reached (*atEnd*()). Such a basic iterator is termed a *ForwardIterator*, other kinds of iterator are:

- *Bidirectional iterators*, supporting additional *previous*() and *atStart*() operations.
- *Random access iterators*, supporting access to positions via an index, these iterators provide operations such as *get*(i : *int*).

The general iteration algorithm for a forward iterator *it* is:

```
it.start() ;
while not(it.atEnd())
do
  ( ... process it.element() ...;
    it.next()
  )
```

An unusual but useful application of iterators is to iterate over collections which are too large to hold in memory at once. The iterator instead generates new elements of the collections as needed. For example, all permutations of a large collection (for a collection of size n, the number of permutations is $n!$), or all subsets (2^n subsets for a collection of size n). Figure 5.8 shows examples of such iterators.

The definitions of *next* for *PermutationIterator* and *SubsetIterator* are based on the *increment* operation for a lexicographic ordering. For example, for an alphabet a, b, c, the least element of the ordering is the empty string "", then successive elements are "a", "b", "c", "aa", "ab", etc. The permutations of the alphabet sequence are all the words with the same length as the alphabet and containing no duplicates. The first permutation is *alphabet*, whilst *next* for a permutation iterator starts at the current permutation *word* and successively applies *increment* until the next permutation is obtained. The last permutation is *alphabet.reverse*.

The *PermutationIterator* can be used as follows in Java:

```
public static void main(String[] args)
{ PermutationIterator p = new PermutationIterator();
  Vector v = new Vector();
  v.add("a"); v.add("b"); v.add("c");
  p.setbasesq(v);
  p.init();
  p.start();
  List w = p.element();
  System.out.println(p.element());

  while (! p.atEnd() )
```

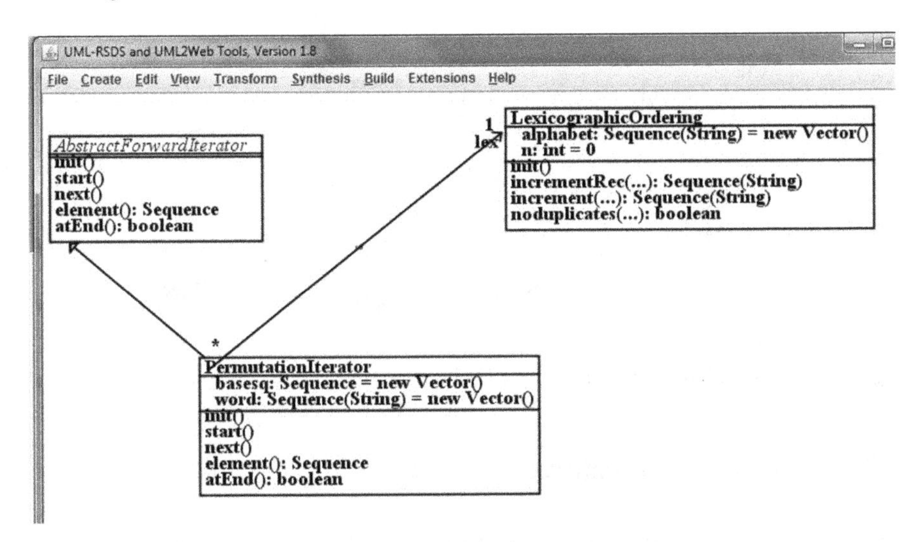

Fig. 5.8 Iterators for large collections

```
    { p.next();
      System.out.println(p.element());
    }
  }
```

5.6.4 *Proxy Pattern*

This pattern addresses the situation where an application needs to communicate with remote objects (e.g., on another computer). The idea of the pattern is to define a local proxy for the remote object, which is more convenient for the application to use. The knowledge of how to invoke the remote object is then separated from the main application out into the proxy: the proxy acts as an intermediary interface or facade for the remote object (Fig. 5.9).

For example, a trading application could be organised in this manner, with a FIX (Financial information exchange) protocol engine being used as a proxy for a remote exchange (Chap. 6).

The application requests services from the proxy object, e.g., to buy a quantity of a share. The proxy handles the details of how the request is implemented, e.g., by constructing a FIX message and communicating with an exchange to make the purchase, using the FIX protocol.

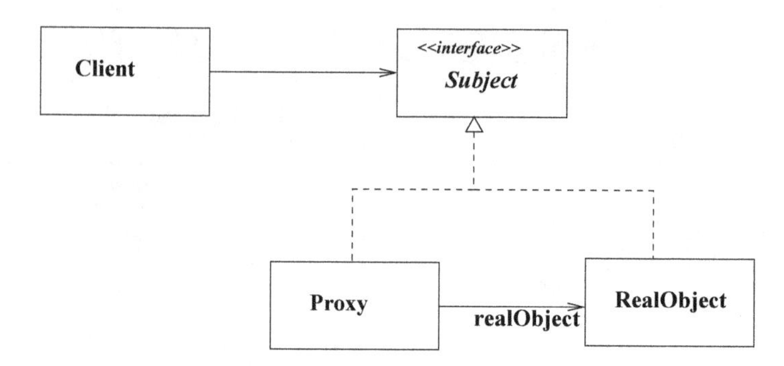

Fig. 5.9 Structure of proxy pattern

5.7 Software Architectures

The large-scale design of a software application involves the determination of which software components are needed, and the organisation of the component interactions in order to achieve specified system requirements.

Components should be:

- Subsystems and modules with a coherent and well-defined purpose, satisfying functional cohesion.
- A group of closely related classes, or a single class.

Modules are highly-cohesive components which carry out a specific set of functions. For example, managing account creation, deletion and customer-account dependencies in a bank. A separate module may be concerned with managing account transactions, since these functionalities are relatively independent of each other.

Subsystems are groups of modules and may be organised on the basis of a general role within the application. For example, a UI or client subsystem is concerned with receiving information from and presenting information to users, whilst a functional core or business functions subsystem contains the modules which define the core business logic of the application. A data repository or resource subsystem contains modules which manage data or links to external resources. Such subsystems are usually referred to as 'tiers' since they are organised as a stack of layers or tiers, with communication only between adjacent layers.

A number of possible architectures can be appropriate for financial applications: client-server (2 tier, with the business and resource tiers combined), client-business-resource (3-tier), or more complex architectures such as an enterprise information systems (EIS) architecture involving additional intermediate tiers between the client, business and resource tiers.

We will describe architectures using *architecture diagrams*, which represent components (subsystems or modules) as rectangles, with nesting used to indicate packaging of components within others. An arrow from component X to component Y means that X *depends on* Y: it invokes operations of Y, or it refers to data of Y. X

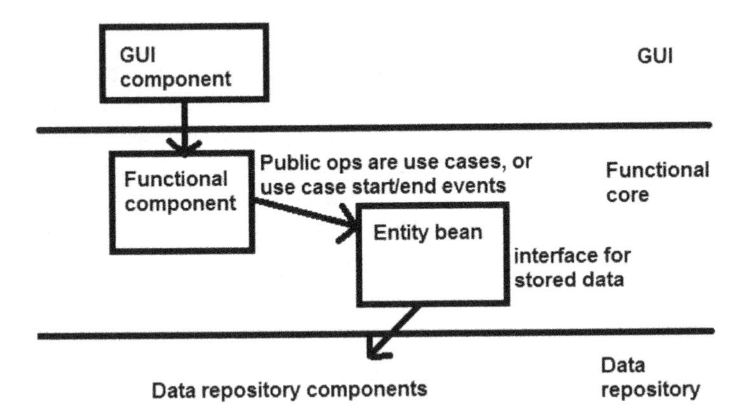

Fig. 5.10 General 3-tier application architecture

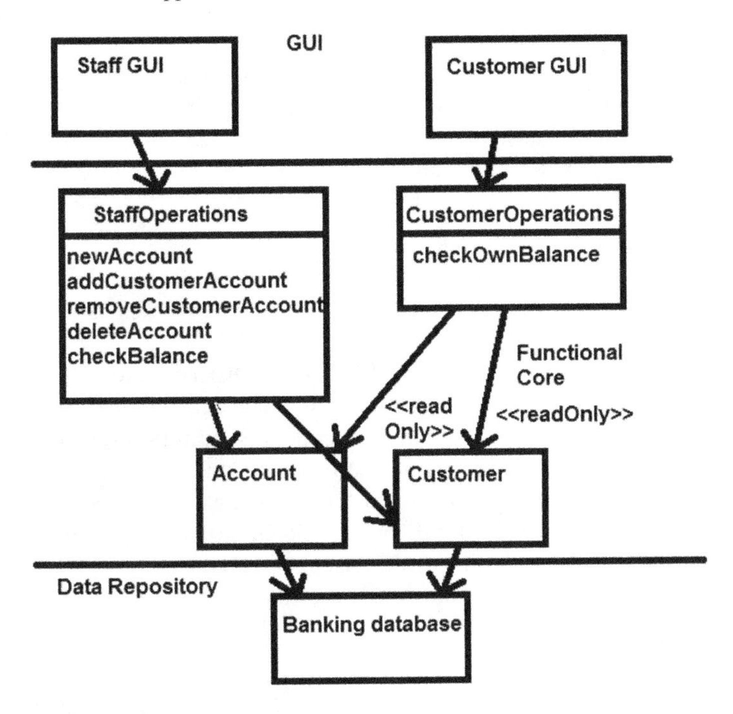

Fig. 5.11 Banking system architecture

is termed a *client* of Y, and Y is a *supplier* of X. The arrow can have a stereotype $\ll$ *readOnly* $\gg$ to indicate that the client does not modify any of the supplier data. The operations of a module (the services it offers to clients) can be listed in the module rectangle. Figure 5.10 shows a typical 3-tier application architecture in this notation.

Within the functional core/business tier there can be a further subdivision of components in which certain components are functionally-oriented and define a set of functions which the business tier provides as services to the higher tier. The functions may coincide with the use cases of the application or with special cases of them. Other components are data-oriented and represent the key business entities. The functional components (also termed *session beans*) co-ordinate the functionalities provided by the data components (or *entity beans*) to carry out their own operations. Business data in the entity beans is persisted by storing it in data resources via the data repository/resource tier. A further example of this architecture is shown in Fig. 5.11.

5.8 QuantLib

QuantLib is an open source C++ library for quantitative finance (www.quantlib.org). It was launched in 2000 and has become widely used in the finance domain. Although companies may have misgivings about using code over which they have no control, open source libraries have advantages over proprietary code (such as Excel) because users can see precisely what computations are being performed. The widespread uptake and usage of libraries can increase their reliability (as for example, with the programming language libraries for C, C++, Java, etc).

QuantLib modules include:

- Numeric types: synonyms such as *Real*, *Time*, *Rate* for *double*.
- Currencies, FX rates: including predefined national currencies and functions for the management of exchange rates.
- Design patterns: templates for Factory, Singleton, Observer, etc.
- Date and time: dates, calendars, day counts and date calculations.
- Math tools: random number generators, root finders, optimisation algorithms.
- Monte Carlo: a framework for Monte Carlo simulations.
- Cash flows: classes and functions for cash flows and cash flow sequences.
- Term structures: a framework for defining yield curves/term structures.
- Financial instruments: including a wide range of different contract models.
- Pricing engines: pricing algorithms for many kinds of financial instrument.
- Volatility models.
- Stochastic processes.

QuantLib therefore forms a natural basis for a DSL for quantitative finance, and in this book we have endeavoured to define our models in a form that is consistent with QuantLib.

5.8.1 Design Patterns in QuantLib

QuantLib uses the following design patterns:

- *Factory*: a class which produces instances of other classes. A specific example is a factory that creates different varieties of financial instruments.

- *Singleton*: used for repositories of data that are used in several places in the code. E.g., a repository holding market data points for interest rates, or a repository of exchange rates between currencies.
  Repository :: *instance*() returns the instance of singleton class *Repository*.
- *Observer* is used in any situation where some objects may need to be notified when another object changes state.
  For example, a bond valuation object that depends upon a yield curve: the valuation must be recalculated if the yield curve changes. Therefore the valuation object must be an observer for the yield curve object.
  The QuantLib *Quote* class is both an Observable (a Subject), and an Observer (a View).

Summary

In this chapter we have considered software design approaches for financial systems, using UML and agile MBD. We have described reuse techniques, design patterns, and given an overview of the QuantLib finance library.

Exercises

1. Specify a library function $coprime(x : int, y : int) : boolean$ which returns *true* if x and y are positive integers which are co-prime (they have no common divisor except 1). E.g., 21 and 25 are co-prime. Define both a postcondition and an activity for the operation.

2. Use the bootstrap procedure to find 4 year term and 5 year term interest rates, if (i) the known 1, 2 and 3 year term rates are 1.5, 1.8, 1.7%, (ii) there is a 4-year 2% annual coupon bond with price 103 and (iii) a 5-year 2% annual coupon bond with price 105. All of these bonds are from the same issuer and settlement date.

3. Define an *update*() operation for an *Investment* (Fig. 5.2) considered as an observer of a *YieldCurve* object which has a query operation $yield(t : double) : double$ to return the yield for a given maturity t.

4. Examine the FX code in QuantLib and define a corresponding class diagram of the concepts and main operations involved in representing and processing FX.

References

1. H. Alfraihi, K. Lano et al., The impact of integrating Agile software development and MDD: a comparative case study, SAM (2018)
2. W. Cunningham, The WyCash Portfolio Management System. Addendum OOPSLA **92**, 29–30 (1992)
3. X. He, P. Avgeriou, P. Liang, Z. Li, Technical debt in MDE: A case study on GMF/EMF-based projects, MODELS (2016)
4. G. Campbell, P. Papapetrou, *SonarQube in Action* (Manning Publications Co, 2013)
5. J. Letouzey, The SQALE Method: definition document version 1.1 (2016)

Chapter 6
Trading and Analytics Technologies

In this chapter we describe technologies to support financial trading and data analysis.

- Trading technologies: the FIX protocol and FIXML
- Data analytics technologies: NoSQL databases, HBase, Map/Reduce, Tensorflow
- Case studies of data analysis: share volatility and technical analysis.

6.1 Trading Technologies: FIX and FIXML

FIX is the *Financial Information eXchange protocol*, an open and de-facto standard for financial information exchange, e.g., to make a trade. FIX defines a message format and a communication model by which the parties to a financial transaction can send and receive messages. It is platform independent, the messages are plain text and can be generated and processed by programs in any programming language. FIX originated in the mid 90's from the need to standardise electronic trading and order management formats.

The main uses of FIX are to support electronic trading, involving exchanges, brokers and other market participants; share and fixed income trading, and streaming multicast of financial data via FAST (FIX Adapted for STreaming).

The basic text format of a message is a sequence of `field=value` bindings, separated by a delimiter character (ASCII code 1, the SOH character). The fields are integers. Numbers from 0 to 4999 have predefined meanings, numbers from 5000 to 9999 are available for application-specific extensions. A header segment identifies the FIX version (field 8), the message body size (field 9), message sequence number within a session (field 34), the message type (field 35), etc. The body segment contains data of the specific transaction, such as an account identifier (field 1), price (field 44), etc. Finally a message trailer contains a checksum (field 10).

K. Lano and H. Haughton, *Financial Software Engineering*, Undergraduate Topics in Computer Science, https://doi.org/10.1007/978-3-030-14050-2_6

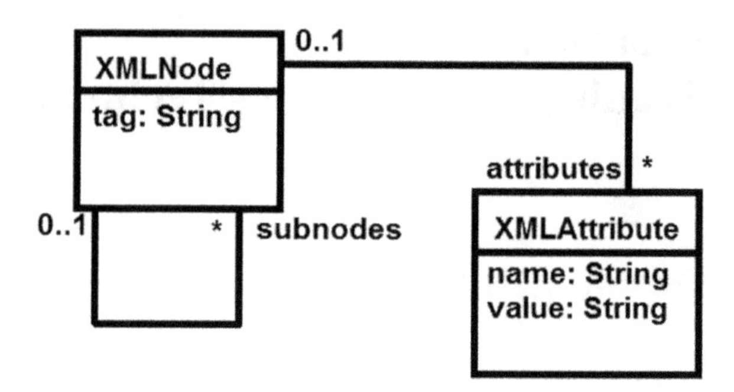

Fig. 6.1 XML metamodel

Messages can be administration messages, such as session establishment and termination messages, resend requests, etc., or messages enacting financial process steps, such as an order creation, position report, etc.

6.1.1 XML

XML is a structured text format used as the basis for HTML and many other description and interchange representations. An XML document is a nested tree structure of nodes with attributes and subnodes. Sometimes XML files are written by hand (e.g., small web pages or configuration files). More often they are generated by software. For example, XML is used for messages and service definitions in the SOAP web service framework.

Figure 6.1 shows a simplified version of the XML metamodel.

XML nodes are named by a *tag* identifying what the node represents:

```
<tag1 .... />
```

This is a node with no subnodes.

Nodes can have attributes, with values:

```
<tag1 att1="val1" att2="val2" />
```

Nodes may be nested to any depth:

```
<tag1 .... >
  <tag2 ... />
  <tag3 ... />
</tag1>
```

In this case simple nodes with tags *tag*2 and *tag*3 are nested in a composite node with *tag*1.

As an example, the XML data

```
<car make="XJ6" colour="silver" manufacturer="Jaguar">
```

```
  <engine capacity="31" />
  <DVLARecord status="OffRoad" date="20120101"/>
</car>
```

represents a particular car.

6.1.2 FIXML

FIX messages can be expressed either in the basic character format or in an XML format, called FIXML. The FIXML format is defined at http://fixwiki.org/fixwiki/, and http://www.fixtrading.org. FIXML files are produced and consumed by software, e.g., trading applications.

An example of a FIXML file is a simple order message:

```
<?xml version="1.0" encoding="ASCII"?>
<FIXML>
  <Order ClOrdID="123456" Side="2"
             TransactTm="2001-11-17T09:30:47-05:00" OrdTyp="2"
         Px="93.25" Acct="26522154">
    <Hdr Snt="2001-11-17T09:30:47-05:00"
             PosDup="N" PosRsnd="N" SeqNum="521">
      <Sndr ID="AFUNDMGR"/>
      <Tgt ID="ABROKER"/>
    </Hdr>
    <Instrmt Sym="IBM" ID="459200101" IDSrc="1"/>
    <OrdQty Qty="1000"/>
  </Order>
</FIXML>
```

The first line indicates the version of XML that is being used. The FIXML tag identifies that the data is FIXML format. The file content defines an *Order* with nested header *Hdr*, instrument *Instrmt* and *OrdQty* subnodes. The header has nested sender and target subnodes. The names of FIX fields are used instead of numbers in the basic text format. Thus *Acct* is used instead of the number 1 to mark the account.

A more complex example is a Position Report message:

```
<?xml version="1.0" encoding="ASCII"?>
<FIXML>
<PosRpt RptID="541386431" Rslt="0"
  BizDt="2003-09-10T00:00:00" Acct="1" AcctTyp="1"
  SetPx="0.00" SetPxTyp="1" PriSetPx="0.00" ReqTyp="0" Ccy="USD">
<Hdr Snt="2001-12-17T09:30:47-05:00" PosDup="N" PosRsnd="N" SeqNum="1002">
<Sndr ID="String" Sub="String" Loc="String"/>
<Tgt ID="String" Sub="String" Loc="String"/>
<OnBhlfOf ID="String" Sub="String" Loc="String"/>
<DlvrTo ID="String" Sub="String" Loc="String"/>
</Hdr>
<Pty ID="OCC" R="21"/>
<Pty ID="99999" R="4"/>
<Pty ID="C" R="38">
<Sub ID="ZZZ" Typ="2"/>
```

```
</Pty>
<Qty Typ="SOD" Long="35" Short="0"/>
 <Qty Typ="FIN" Long="20" Short="10"/>
 <Qty Typ="IAS" Long="10"/>
<Amt Typ="FMTM" Amt="0.00"/>
<Instrmt Sym="AOL" ID="KW" IDSrc="J" CFI="OCASPS" MMY="20031122"
 Mat="2003-11-22T00:00:00" Strk="47.50" StrkCcy="USD" Mult="100"/>
</PosRpt>
</FIXML>
```

In this case there are multiple subnodes with the same tag (Pty and Qty) for different contracts within the position.

Other sample FIXML messages can be found at http://fixwiki.org/fixwiki/.

6.1.3 FIX Engines

A FIX engine is a software component which creates and processes FIX messages, either in the basic text format or in FIXML or in both. It can be called from an application via an API in order to carry out trading commands via the FIX protocol. There are FIX engines for many different platforms and programming environments. It is also possible (but time-consuming) to implement your own engine.

FIX engines act as endpoints of a FIX communication channel between financial applications (Fig. 6.2). They manage the administration of the communication and enforce the communication protocol including timeouts and error detection and recovery.

A communication session is initiated by the client via a *logon* message to the server. The server receives the request, and after validation, establishes the connection. The client can end a session by sending a *logout* message. Individual messages in a session are identified by a sequence number (*SeqNum* field in the above examples). This permits detection of messages which arrive out of order, lost or invalid

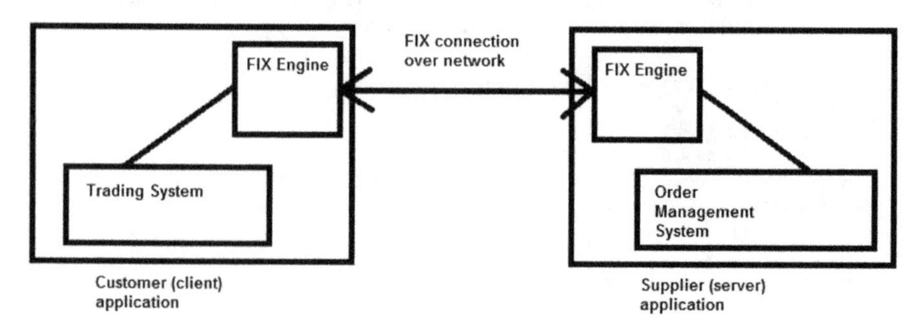

Fig. 6.2 Role of FIX engines

messages, etc. A special kind of message (heartbeat messages) are sent in both directions at regular intervals in order to maintain the connection. The session may be closed if no message of any kind is received for 2 times the heartbeat interval.

6.2 Data Analytics Technologies

The representation and processing of 'big data' requires new database and analysis technologies, in particular alternatives to relational databases, such as column-oriented databases, and distributed large-scale parallel processing models such as Map/Reduce.

6.2.1 Data Storage

Relational databases have been the mainstream database technology since the 1970s. They satisfy the standard properties of supporting CRUD operations (Create, Read, Update, Delete), and their transactions have the ACID (Atomicity, Consistency, Isolation, Durability) properties. Relational databases use *normalised* tables to reduce data redundancy, and the SQL query language is used to update and search data.

While relational databases have been very successful for many tasks, they have limitations for storage of very large (gigabyte and upwards) data sets and efficient processing of these. Thus various forms of *NoSQL databases* have been devised. These also support CRUD operations but relax some ACID properties. Data may not be normalised, which may improve efficiency but results in data duplication, taking advantage of the declining costs of data storage. Instead of SQL, NoSQL databases use specialised APIs to read and update data.

Some example NoSQL data storage technologies are Bigtable and HBase. These use column-oriented data models for efficient storage of big data. Bigtable was developed at Google for applications such as web indexing, Google Earth, Google Finance, etc.

The range of NoSQL data models include:

- Key-value stores such as Oracle Coherence
- Column-oriented databases such as Bigtable and HBase
- Document databases such as MongoDB
- Text-search datastores such as Apache Lucene

In the column-oriented data model, a data store consists of a set of tables, and the table rows have the structure

```
row key, column family 1, ..., column family n
```

where `row key` is a string, unique for each row, and each `column family` consists of one or more related columns of data cells. The rows are stored in lexicographical order of the keys,[1] and each cell can store multiple versions of data, each version is timestamped.

An example of this form of data storage is the Bigtable store for Google Analytics, which records website statistics (visits per day to different websites, details of each visit, etc). A table records session data, the row keys of the session table are of the form:

```
website name + date/time of session start
```

Thus all session data for visits to one website are stored contiguously, and the records are stored in chronological order.

The Apache HBase database (hbase.apache.org) is a column-oriented, key-value database. Each data item in a table is addressed by a *row key*, a *column family*, and *column name* as for Bigtable. There can be multiple timestamped versions of items. Rows are stored in lexicographic order of row key. The choice of key structure is important to ensure efficiency. Generally we should choose the key so that (i) data likely to be frequently accessed together are close in the table; (ii) data accesses are otherwise distributed across a wide range of the table, to avoid overloading one node in the distributed data store. Physically, large tables can be split into disjoint blocks of rows, stored on separate computers.

An HBase example for finance is share data analysis. Data on share trading on particular exchanges can be obtained from sites such as AlphaVantage (alphavantage.co) or IEX Trading (https://iextrading.com) with a format such as

```
Symb, date, opening price, high, low, closing price, volume
```

where *Symb* is an alphanumeric company symbol, e.g., "IBM". *opening* and *closing* prices are the prices at the start and end of the trading day *date*, and *volume* is the trading volume of shares of *Symb* on *date*. *high* and *low* are the maximum and minimum prices reached by *Symb* on *date*. *date* typically has a format such as day/month/year (2 digits for the day, followed by 2 for month then 4 for year).

If we want to ensure that all data for one symbol is stored together, in chronological order, we need to define the keys as:

```
Symb:year:month:day
```

This means that the rows (e.g., for IBM) are in the following order:

```
IBM:2017:12:28 ....
IBM:2017:12:29 ....
IBM:2018:01:02 ....
```

[1]Recall that "a0" < "a1" in lexicographic order, and "aa" < "ab", etc.

This ordering ensures that data for the same share for successive trading days are always stored in successive rows. If we had instead used the day/month/year format, data for the same day in different months and years would be stored contiguously.

Our key makes sense if accesses to nearby dates of one symbol frequently occur within one operation (e.g., to examine price trends for the symbol over a time period). The same dates for different shares are distributed across the table.

The column organisation can consist of a column family *prices* with four columns *opening*, *closing*, *high*, *low*, and a column family *volume* with one column.

There are APIs for HBase in different languages. The Java API uses the following commands to create a table and add a column family (table, row key and column names are byte arrays):

```
HTableDescriptor table = new HTableDescriptor(
        TableName.valueOf("ShareTable".getBytes()));
table.addFamily(new HColumnDescriptor("prices".getBytes()));
```

To insert a row with key k, column family *cf* and columns $c1$, $c2$, use:

```
Put p = new Put(k);
p.addColumn(cf,c1,data1);
p.addColumn(cf,c2,data2);
table.put(p); // or a list of Put's
```

The *get* method obtains single rows or cells:

```
Get g = new Get(k);
g.addColumn(cf,c1);
Result r = table.get(g);
if (r.isEmpty())
{ System.err.println(
    "No data for row " + k + " cell " + cf + ":" + c1); }
else
{ System.out.println("" + r.value()); }
```

The *scan* command obtains sets of rows. To scan *rn* rows starting from k we can write:

```
Scan scan = new Scan();
scan.setStartRow(k);
scan.setCaching(rn);
ResultScanner results = table.getScanner(scan);
for (Result r : results)
{ // do something with r
}
```

It is also possible to filter results in the scan.

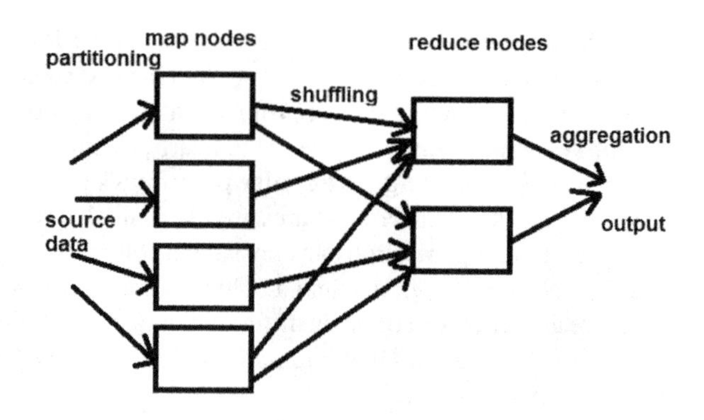

Fig. 6.3 Map/Reduce processing model

6.2.2 *Map/Reduce*

Map/Reduce is a framework for the distributed processing of large datasets. The process partitions the source dataset into blocks, which are separately processed by the *map* function, possibly on separate computers. The result sets are combined via the *reduce* function to produce an overall result. This approach is suitable for analysis of large static datasets (not streams). Map/Reduce implementations incorporate reliability via distribution and the reassignment of work from failed processors.

A Map/Reduce platform provides:

- Input data partitioning and assignment of partitions for *map* processors
- Partitioning and shuffling of *map* output to the *reduce* processors
- Crash recovery.

Figure 6.3 shows the general structure of Map/Reduce processing.

For a particular analysis task, a developer writes appropriate *map* and *reduce* functions of the following forms:

- $map : KeyT1 \times ValueT1 \rightarrow Sequence(KeyT2 \times ValueT2)$
  map takes pairs $(k1, v1)$ and produces lists of $(k2, v2)$ pairs.
- $reduce : KeyT2 \times Sequence(ValueT2) \rightarrow Sequence(ValueT3)$
  reduce takes a $(k2, [w1, \ldots, wm])$ pair and produces lists $[u1, \ldots, up]$.

The shuffle step combines *map* results: (k, v) and (k, u) are combined to $(k, [v, u])$, etc.

reduce can be performed in stages, if it is *associative*, i.e., the result of several *reduce* nodes can be fed into a further *reduce*.

As an example, consider the task of counting occurrences of different company symbols in a large collection of FIXML Order message files m.

- $map(m.name, m.data)$ is defined to produce $[(s, 1)]$ if $m.data$ is an Order and $Sym = s$ occurs in m.data, otherwise it produces the empty sequence []
- Shuffle directs aggregated data (s, sq) for a given s to a *reduce* node for s
- $reduce(s, counts)$ produces $[(s, \Sigma counts)]$.

The aggregated final result is a list of symbols with their counts.

In this case *reduce* is associative: different count sums for the same s could be fed into a further *reduce*.

General queries can be implemented using Map/Reduce, for example an OCL expression

$$data \rightarrow select(x \mid P) \rightarrow collect(e) \rightarrow r()$$

with r associative can be implemented as follows:

- Split *data* into partitions based on some *key*1 of its elements
- *map* produces e-sequences from the x satisfying P
- Results are allocated to *reduce* based on a *key*2 of the e-values
- *reduce* applies r.

6.2.3 Apache Hadoop

This is an open-source framework for distributed storage and processing of Big data sets (hadoop.apache.org). It consists of:

- The Hadoop distributed file system (HDFS)
- Hadoop MapReduce implements Map/Reduce using HDFS.

Hadoop also supports HBase and other packages.

6.2.4 TensorFlow

TensorFlow is an open-source platform for machine learning, it supports the construction of neural networks for learning data classification rules (https://www.tensorflow.org). TensorFlow processes data as multi-dimensional arrays/matrices, termed *tensors*. A training phase builds the classifier using existing data with known classifications. An example application of Tensorflow could be to learn trading strategies based on historical share data and trading indicators.

6.3 Case Study: Share Price Volatility

The *volatility* of a share or other financial asset is a measure of how variable its price or value is over a period of time. This can be used to assess the risk of investing in the asset (the higher the volatility the greater the risk of losses) and it is also important in computing the value of a derivative security based on the asset. The volatility is conventionally denoted by σ, and represents the standard deviation of the change

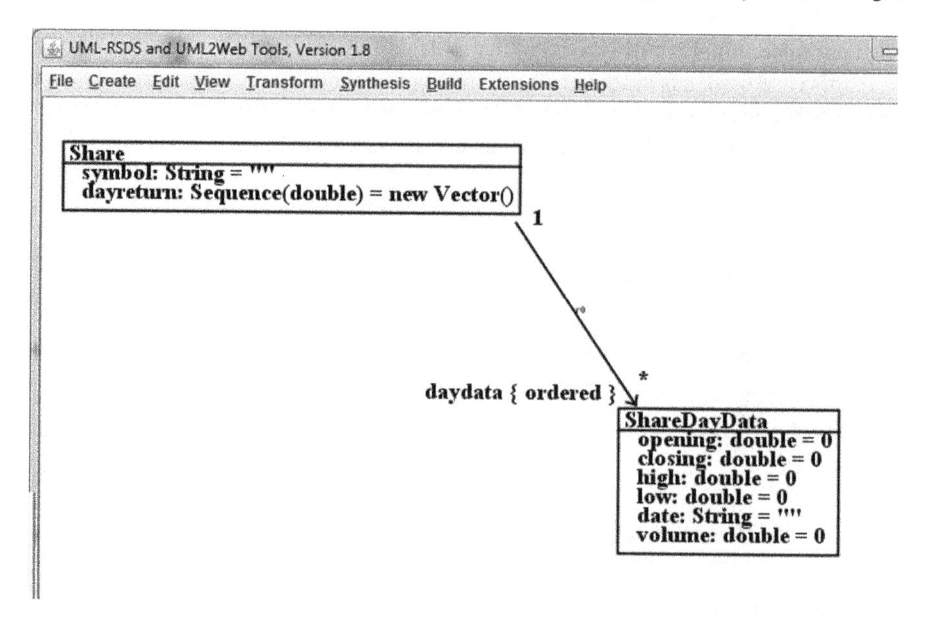

Fig. 6.4 Share model

in the asset price over a period of time, usually one year. More precisely, σ is the standard deviation of the continuously compounded return of the asset over one year.

One way to estimate σ for a particular share, e.g., for IBM, is to sample closing prices $closing_i$, $i = 1$ to n, of the share over a relevant period (e.g., the previous 90 trading days), compute the proportional changes in price $x_i = closing_i/closing_{i-1}$ for $i = 2$ to n, and compute the standard deviation s of the $log(x_i)$. The volatility is then $s/\sqrt{T}$ where T is the length of the sampled time period in years.

We can express this procedure in terms of the share model of Fig. 6.4, where *daydata* holds a sequence of the daily price and volume information for the share, assumed to contain sufficient information for analysis. Physically, this could be stored in an HBase table.

As a financial procedure, the following steps are required:

1. Get the closing prices for the last 90 trading days (assuming $daydata.size > 90$):

$$days90 = daydata.subrange(daydata.size - 89, daydata.size)$$

   for an auxiliary sequence variable *days*90.

2. Compute the daily returns:

$$dayreturn[i-1] = (days90[i].closing/days90[i-1].closing) \rightarrow log()$$

   for $i = 2$ to 90.

3. Compute the standard deviation of the returns, using a library function:

$$s = StatLib.standardDeviation(dayreturn)$$

4. Set the volatility of the share to be $s/(300.0/90.0)\rightarrow sqrt()$, assuming a 300 day trading year.

$$volatility = StatLib.standardDeviation(dayreturn)/(10.0/3.0)\rightarrow sqrt()$$

In formal OCL notation, these steps become the postconditions of a use case *computeVolatility*:

$$\begin{array}{l} Share :: \\ \quad days90 = daydata.subrange(daydata.size - 89, daydata.size) \Rightarrow \\ \qquad Integer.subrange(2, 90)\rightarrow forAll(i \mid \\ \qquad\quad (days90[i].closing/days90[i-1].closing)\rightarrow log() : dayreturn) \end{array}$$

$$\begin{array}{l} Share :: \\ \quad volatility = StatLib.standardDeviation(dayreturn)/(10.0/3.0)\rightarrow sqrt() \end{array}$$

6.4 Case Study: Technical Analysis of Share Prices

Technical analysis involves the analysis of share price data over past periods, with the aim of detecting trends and predicting future price movements, in order to guide investment decisions. For example, if there has been a consistent downward trend in a share price, it is useful to be able to recognise a point where the trend ceases and could reverse: this could be a opportunity to invest in the share, or at least to cease selling it. Conversely if a trend of price increases is coming to an end, this would be an opportunity to sell the share. Software support is critical to technical analysis, and this involves the computation of different technical indicators such as various forms of moving averages of share prices over an interval of time, these indicators can either be for the direct use of a technical analyst, or they can be used as input to machine learning algorithms such as neural nets.

The basis of several technical indicators is the concept of a *moving average*: a computation of the average price of a share over a preceding period (e.g., the previous 26 trading days). As time passes, the oldest price in the series is replaced by the newest, and the average is recalculated.

Given the model of Fig. 6.4, a 26-day average of closing prices is:

$$SMA(26)(n) = (\Sigma_{i=n-25}^{n} daydata[i].closing)/26$$

Recall that sequences are indexed starting from 1.

In OCL, the definition is:

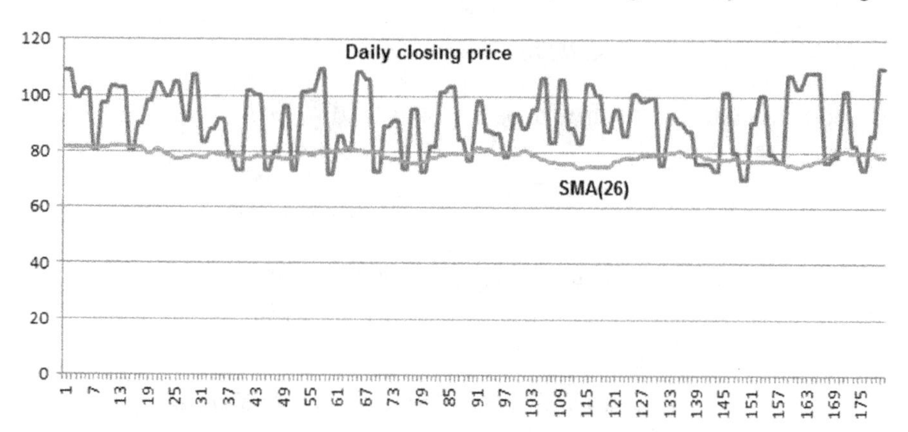

Fig. 6.5 Daily price and SMA graph for a share

$$SMA(26)(n) = \\ (daydata.subrange(n - 25, n) \rightarrow collect(closing) \rightarrow sum())/26$$

for any n from 26 up to *daydata.size*.

The 26-day moving average sequence *sma* can be maintained together with *daydata*:

```
Share::
  adddata(d : ShareDayData)
  pre: daydata.size > 25 & sma.size > 0
  post:
    n = daydata@pre.size &
    smad = sma@pre.last + d.closing/26 -
        daydata@pre[n-25].closing/26 &
    daydata->includes(d) &
    sma->includes(smad)
```

Figure 6.5 shows a graph of daily share price data (upper line) and the corresponding SMA(26) line. The SMA line smoothes out the variations in the daily line. Points where the daily line crosses from below the average line to above it are potential indicators to buy the share.

A simple average has defects, in particular, older information is treated equally to recent information. An alternative is an *exponential moving average* (EMA), which weights more recent information more strongly than older data:

$$EMA(26)(n) = \Sigma_{i=n-25}^{n} daydata[i].closing * \alpha * (1 - \alpha)^{(n-i)}$$

That is:

$$EMA(26)(n) = Integer.Sum(n - 25, n, i, \\ daydata[i].closing * \alpha * ((1 - \alpha) \rightarrow pow(n - i)))$$

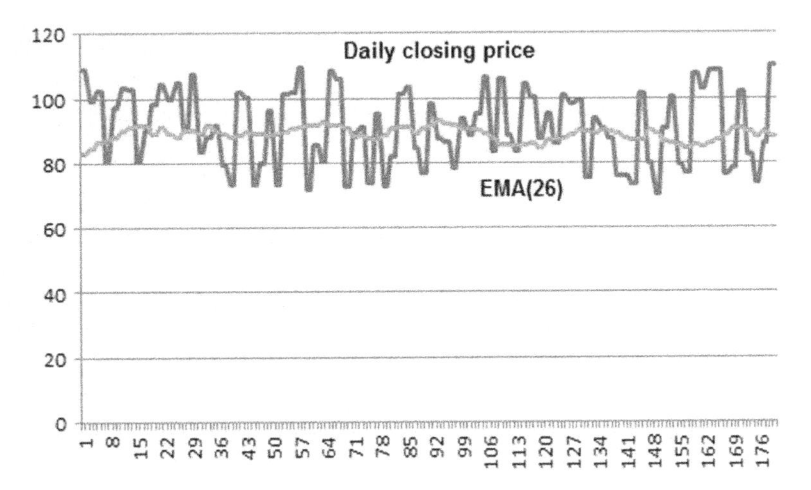

Fig. 6.6 Daily price and EMA graph for a share

where $n \geq 26$ and $n \leq daydata.size$, for a 26-day exponential moving average with decay factor α. A larger decay factor weights more recent data more heavily. A common choice of α is $\frac{2}{N+1}$ where N is the term of the EMA. For $N = 26, \alpha = 0.074$ approximately.

Again, the EMA can be maintained together with the daily data:

```
Share::
  adddata(d : ShareDayData)
  pre: true
  post:
    emalast = (if ema.size > 0 then ema.last else 100 endif) &
    emad = emalast + alpha*(d.closing - emalast) &
    daydata->includes(d) &
    ema->includes(emad)
```

Figure 6.6 shows a graph of daily share price data and the corresponding EMA(26) line. Again, points where the daily line crosses from below the average line to above it are potential indicators to buy the share.

From the EMA, further indicators can be derived. A popular indicator is the *moving average convergence/divergence* (MACD). The MACD is the difference between a short-term EMA, e.g., for 12 days, and a longer-term EMA, e.g., for 26 days:

$$MACD(12, 26)(n) = EMA(12)(n) - EMA(26)(n)$$

The short-term EMA reflects recent share price behaviour more strongly than does the long-term EMA, so at points when $EMA(12)$ becomes greater than $EMA(26)$, i.e., the short-term graph crosses the long-term from below, there is evidence of an increasing price trend (or 'bullish' sentiment in the market towards the share). This corresponds to the MACD crossing the zero line from negative to positive values.

Conversely if the MACD crosses the zero line from positive to negative, this can indicate a decreasing price trend and 'bearish' sentiment in the market.

An EMA of the MACD is additionally computed, typically on a 9-day basis:

$$MACDAVG(n) \;=\; (EMA(9)(MACD(12, 26)))(n)$$

Finally, the divergence or difference between MACD and MACDAVG is also computed.

A 'signal-line crossover' is a situation when MACD and MACDAVG cross over. This is taken as a buy signal if the MACD becomes greater than the MACDAVG, and as a sell signal if the reverse happens.

Summary

In this chapter we have reviewed some important underlying technologies for finance: the FIX protocol enables transfer of financial data and service requests in a standardised format, either a basic text format or structured XML (FIXML). Data analytics technologies such as NoSQL databases and Map/Reduce enable large scale storage and analysis of financial data.

Exercises

1. Daily share price data is often provided in the form

```
symbol,date,opening,high,low,closing,volume
```

where *symbol* is the stock symbol such as "IBM", *opening*, *closing*, *high* and *low* are the prices for *symbol* on *date*, and *volume* is the amount of trades in *symbol* on *date*.

For example:

```
"IBM", 28/12/2017, 66.32, 66.35, 67.12, 66.25, 9895653
"IBM", 29/12/2017, 66.42, 65.95, 66.92, 65.55, 9011653
"MSFT", 27/12/2017, 85.65, 85.98, 85.215, 85.71, 14662085
```

Define map/reduce processing to obtain (i) for each symbol, a count of the number of dates on which $volume > 300000$ and $closing \geq opening + 1$; (ii) for each symbol, the dates and difference $closing - opening$ for which $closing - opening$ is at its maximum value.

2. For the above share database, describe how Map/Reduce can be used to find, for each share symbol, the date(s) with the highest closing prices for the symbol, for dates with volume at least 100000.

3. Modify the *adddata* operations above so that they return *true* if the *daydata* graph crosses up over the SMA or EMA graphs at the point of the new data, and *false* otherwise.

4. In the graph of Fig. 6.5, how good a predictor of increased share prices is the 'price crossing SMA line upwards' indicator? How many times after such an event is the price higher 5 days after the event compared to the price at the event? For what percentage of the events is this true?

5. What specific OCL functions r can be used in expressions

$$data \rightarrow select(x \mid P) \rightarrow collect(e) \rightarrow r()$$

computable using Map/Reduce?

Chapter 7
Software Modernisation and Re-Engineering

In this chapter we consider the use of UML and MBD for software modernisation and re-engineering of applications. We consider two case studies:

- Matlab to C# migration case study
- Yield curve estimation

7.1 Software Modernisation

There is a substantial demand in industry for the modernisation of existing applications, including: (i) updating applications to support new modes of interaction such as web services or cloud-based provision; (ii) migrating legacy applications from outdated/unsupported technologies to modern technologies; (iii) transforming applications to meet the requirements of new regulations; (iv) re-architecting applications to improve their fit to a business process.

An example of (ii) is the analysis of a 100 million LOC banking system in COBOL, and the re-engineering of this to 3 million LOC C# [1]. An example of (iv) is the modernisation of a financial data management process described in [2].

Reverse-engineering is the process of extracting information from an existing application in order to obtain a specification and other documentation that describe its functionality and other properties. Re-engineering involves reverse-engineering to obtain a specification that can be used as the basis of a specification for a modernised version of the application. Migration of the specification may be carried out to restructure and reorganise the application data and functionality into a form that is more suitable for the new platform/environment. Then forward engineering is applied to construct a new application version from the extracted specification. Reverse and re-engineering are typically a combination of manual and tool-supported processes.

K. Lano and H. Haughton, *Financial Software Engineering*, Undergraduate Topics in Computer Science, https://doi.org/10.1007/978-3-030-14050-2_7

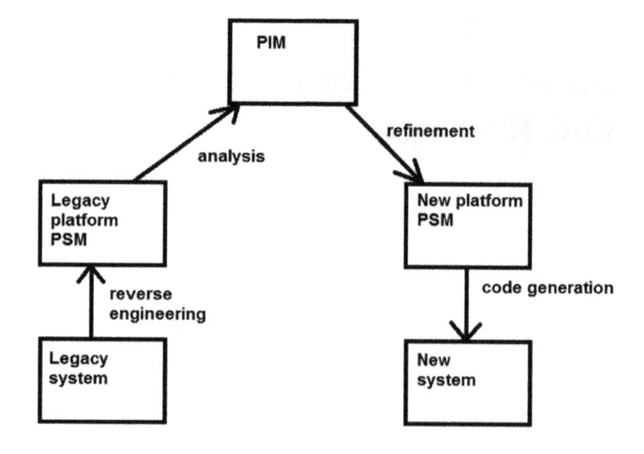

Fig. 7.1 General model-driven re-engineering process

7.1.1 *Model-Driven Modernisation (MDM)*

MDM is defined as a process in which software models are used to support the re-engineering of applications. For example, the OMG's *Architecture driven modernization* envisages the use of tools to analyse and transform legacy components to high-level models, restructuring of these models, and regeneration of a modernised system from the models. The REMICS MDM migration method (www.remics.eu) has the stages: Requirements and feasibility; Recover; Migrate; Validation; Supervise; Interoperability. The method is oriented to modernisation of applications for service-oriented architectures (SOA) and the Cloud.

In general, the MDM re-engineering process (Fig. 7.1) involves:

- Using transformations to analyse an existing system, and to abstract out key information (reverse-engineering) into a platform-specific model (PSM)
- Extracting key business entities and rules from the PSM and expressing them in a platform-independent model (PIM)
- Using transformations to generate a new system/version on a new platform based on the abstracted data and logic.

7.2 Case Study: Matlab to C#

This project aimed to extract functionality from an existing large library of Matlab financial/statistical routines, to rationalise this, and then to re-engineer it to C# code. The business motivation was to remove the dependence of important business services on the poorly-understood and costly to maintain legacy code. In addition, it was intended to move the services to a modern software platform with improved interoperability.

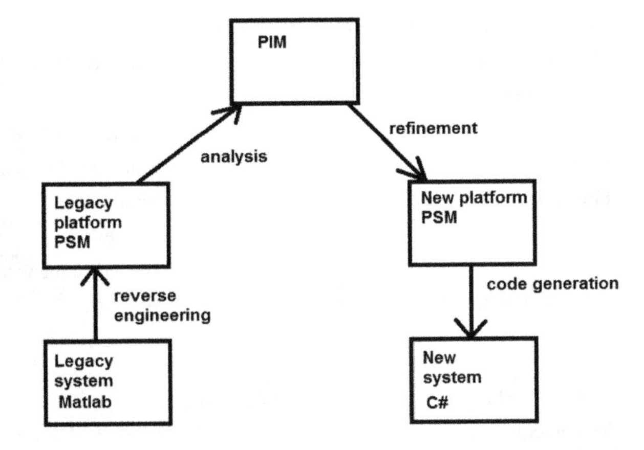

Fig. 7.2 Matlab to C# re-engineering process

A proposed approach for this project was to abstract the application to a platform-independent specification in UML-RSDS, in order to provide high-level specifications of the existing code in the context of relevant financial theory (no specification of the Matlab code existed), and to have some assurance that the C# version has the same functionality as the original (Fig. 7.2). The UML-RSDS representation also supports the use of refactoring and design patterns to improve the system structure. The legacy code architecture is preserved in general, unless it is clearly faulty or inappropriate for the modernised system.

Particularly important for this re-engineering project are the *collect* and *select* OCL operators:

- $col \rightarrow collect(x \mid e)$
  The sequence of values $e(x1)$, ..., $e(xn)$ of $e(x)$ for $x1$, ..., xn in *col*
- $col \rightarrow select(x \mid P)$
  The subcollection of *col* elements x which satisfy P.

For example:

$$Set\{1, 2, 3\} \rightarrow collect(i \mid i * i * i) = Sequence\{1, 8, 27\}$$
$$Set\{\text{“}aa\text{”}, \text{“}bgf\text{”}, \text{“”}, \text{“}xxy\text{”}\} \rightarrow select(s \mid s.size > 2) = Set\{\text{“}bfg\text{”}, \text{“}xxy\text{”}\}$$

These are used to define Matlab matrix and vector operations.

Matlab vectors (arrays of single values) are represented as OCL sequences, e.g., `[3, 5, 7]` becomes $Sequence\{3, 5, 7\}$. Matrices are represented as sequences of sequences, e.g., `[3, 5, 7; 2, 1, 1]` is $Sequence\{Sequence\{3, 5, 7\}, Sequence\{2, 1, 1\}\}$. This provides a direct means of formally specifying Matlab operations, for example, the maximum `max(A)` of a numeric matrix is

$$A \rightarrow collect(s \mid s \rightarrow max()) \rightarrow max()$$

Table 7.1 Correspondence of Matlab and UML

Matlab construct	UML-RSDS representation
Script/program	Use case
Code sections	Included use cases/invoked functions
Global variables	Static attributes of *Global* class
Functions	Functions
Vectors	Sequences
Matrices	Nested sequences
Control statements	Activities

Table 7.2 Mapping of Matlab expressions to UML

Matlab construct	UML-RSDS representation
a:b integers	$Integer.subrange(a, b)$
l:s:u doubles	$Real.subrange(l, s, u)$
prod(a:b) a, b integers	$Integer.Prd(a, b, i, i)$
m(i,:) matrix m	$m[i]$ (i'th row)
m(:,j) matrix m	$m \rightarrow collect(i \mid m[i] \rightarrow at(j))$ (j'th column)
m(i,j)	$m[i] \rightarrow at(j)$ i,j element
m(:)	$m \rightarrow flatten()$ concatenate rows
A.*B	Matrix.mmMult(A,B) matrices A, B
A*B	Matrix.matrixProd(A,B) matrices A, B
A == B	Matrix.mmEq(A,B) matrices A, B

That is, for the elements (rows) of A, collect together their maximum elements, and then take the maximum of these maximums.

Tables 7.1 and 7.2 show the correspondence of Matlab elements and UML elements. The use of function references is simulated by defining suitable *XFunction* classes for different function signatures, and using objects of these classes as function references.

Vector operators are defined in the *Sequences* library, and matrix operators in the *Matrix* library. For example:

```
static query ssAdd(s1 : Sequence(double),
                   s2 : Sequence(double)) : Sequence(double)
pre: s1.size = s2.size
post:
  result =
    Integer.subrange(1,s1.size)->collect( i | s1[i] + s2[i] )
```

in *Sequences* creates a new vector as the element-wise sum of two others. A matrix is represented as a sequence of rows: $m : Sequence(Sequence(double))$, with $m[i]$

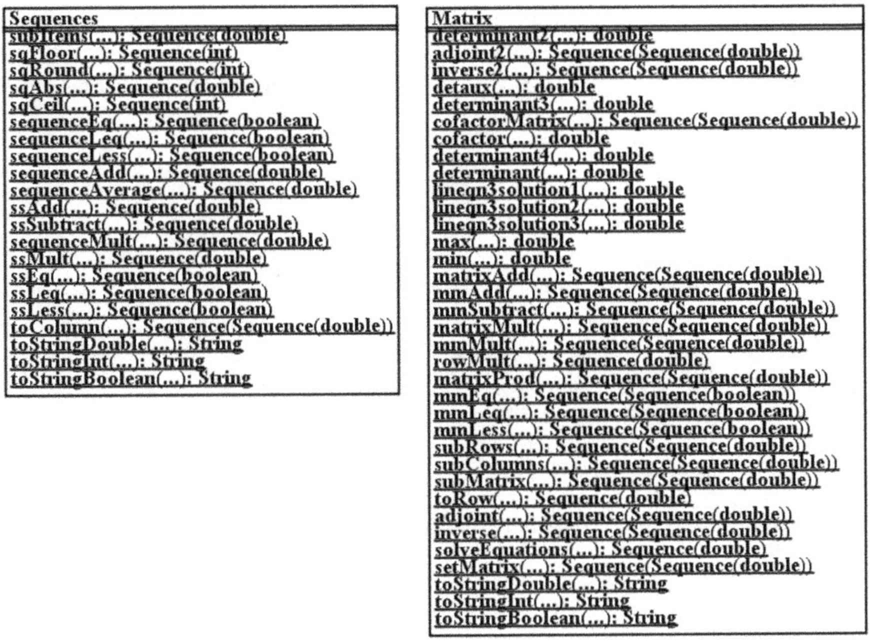

Sequences

subItems(....): Sequence(double)
sqFloor(....): Sequence(int)
sqRound(....): Sequence(int)
sqAbs(....): Sequence(double)
sqCeil(....): Sequence(int)
sequenceEq(....): Sequence(boolean)
sequenceLeq(....): Sequence(boolean)
sequenceLess(....): Sequence(boolean)
sequenceAdd(....): Sequence(double)
sequenceAverage(....): Sequence(double)
ssAdd(....): Sequence(double)
ssSubtract(....): Sequence(double)
sequenceMult(....): Sequence(double)
ssMult(....): Sequence(double)
ssEq(....): Sequence(boolean)
ssLeq(....): Sequence(boolean)
ssLess(....): Sequence(boolean)
toColumn(....): Sequence(Sequence(double))
toStringDouble(....): String
toStringInt(....): String
toStringBoolean(....): String

Matrix

determinant2(....): double
adjoint2(....): Sequence(Sequence(double))
inverse2(....): Sequence(Sequence(double))
detaux(....): double
determinant3(....): double
cofactorMatrix(....): Sequence(Sequence(double))
cofactor(....): double
determinant4(....): double
determinant(....): double
lineqn3solution1(....): double
lineqn3solution2(....): double
lineqn3solution3(....): double
max(....): double
min(....): double
matrixAdd(....): Sequence(Sequence(double))
mmAdd(....): Sequence(Sequence(double))
mmSubtract(....): Sequence(Sequence(double))
matrixMult(....): Sequence(Sequence(double))
mmMult(....): Sequence(Sequence(double))
rowMult(....): Sequence(double)
matrixProd(....): Sequence(Sequence(double))
mmEq(....): Sequence(Sequence(boolean))
mmLeq(....): Sequence(Sequence(boolean))
mmLess(....): Sequence(Sequence(boolean))
subRows(....): Sequence(Sequence(double))
subColumns(....): Sequence(Sequence(double))
subMatrix(....): Sequence(Sequence(double))
toRow(....): Sequence(double)
adjoint(....): Sequence(Sequence(double))
inverse(....): Sequence(Sequence(double))
solveEquations(....): Sequence(double)
setMatrix(....): Sequence(Sequence(double))
toStringDouble(....): String
toStringInt(....): String
toStringBoolean(....): String

Fig. 7.3 Matrix and sequence libraries

as the i'th row. Element-wise addition of matrices uses the *ssAdd* operation to add corresponding rows of the matrices:

```
static query mmAdd(m1 : Sequence(Sequence(double)),
                   m2 : Sequence(Sequence(double))) :
                        Sequence(Sequence(double))
pre: m1.size = m2.size
post:
  result =
      Integer.subrange(1,m1.size)->collect( i |
                              Sequences.ssAdd(m1[i], m2[i] ))
```

Many other vector and matrix operations can be defined in the same manner (Fig. 7.3).

Some example Matlab code is the following section, which forms a Matlab routine to compute a series of put option prices:

```
price = [120.0, 100.0, 102.0, 163.0];
strike = [115.0, 120.0, 95.0, 170.0];
rate = 0.05;
dt = 1.5;
vol = [0.1, 0.2, 0.4, 0.25];
yield = [10.0, 5.0, 4.0, 11.0];
[~, p] = blsprice(price, strike, rate, dt, vol, yield)
```

During analysis we can recognise that *blsprice* is a call to a Black–Scholes European option pricing routine (uk.mathworks.com/help/finance/blsprice.html).

The section is abstracted to a UML-RSDS use case with internal variables *price*, *strike*, etc, and postconditions:

```
::
  price = Sequence{120.0, 100.0, 102.0, 163.0} &
  strike = Sequence{115.0, 120.0, 95.0, 170.0} &
  rate = 0.05 & dt = 1.5 &
  vol = Sequence{0.1, 0.2, 0.4, 0.25} &
  yield = Sequence{10.0, 5.0, 4.0, 11.0}

::
  Integer.subrange(1,4)->forAll( i | p[i] =
        FinLib.euPutOptionPrice(price[i], strike[i],
                                rate, yield[i],
                                dt, vol[i], 0) )
```

This is the formal specification of the section, using the *FinLib* platform-independent financial library.

7.2.1 Yield Curve Estimation

This case study consisted of a large Matlab library to estimate yield curves, using the Nelson-Siegel-Svensson model [3] and variations on this model. It was unclear exactly which estimation algorithms were used, or their validity. Re-engineering involved specification of the required functionality and identification of alternative algorithms/techniques for estimation.

A *yield curve* shows a range of interest rates/yields for investments of different maturities/terms. The yield is the annual rate of return on an investment. Longer loans/investments tend to have higher yields, so a curve normally increases from low terms to higher terms. The yield $y(t)$ for investments of duration t is also referred to as the t-year *spot interest rate*. The investments considered should be zero-coupon bonds, that is, they do not return interest during their term but only at their end. Hence, $y(t)$ is also called the *t-year zero-coupon yield*. Coupon bonds can be converted to equivalent zero-coupon bonds by considering them to have term t equal to their Macaulay duration, with a payment of $value(r)$ on that date (Sect. 4.6).

Only a few time points will have market data or known yields, corresponding to bonds which are available in the bond market, and which are comparable in terms of their origin and risk levels. Yields for other terms are interpolated from the known yields. Different models have been defined for the shape of yield curves. One of the most widely used is the Nelson-Siegal (NS) model for yield curves (Fig. 7.4), which assumes that the yield satisfies a formula

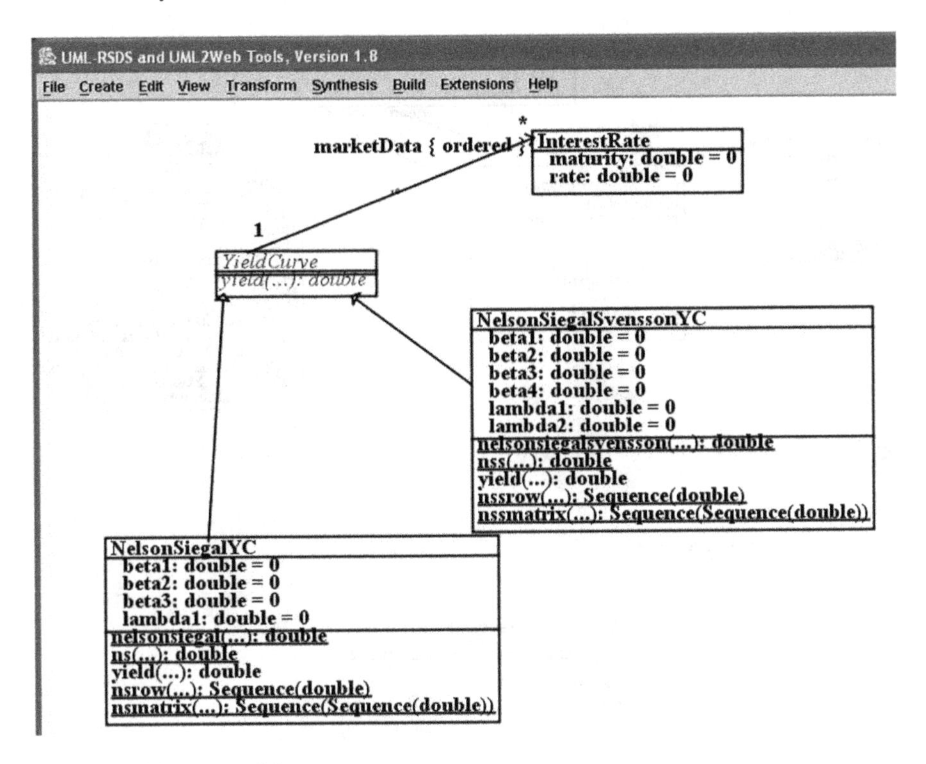

Fig. 7.4 Yield curve models

$$y(t) = \beta_1 + \beta_2 * (1 - exp(-t/\lambda_1))/(t/\lambda_1) + \\ \beta_3 * ((1 - exp(-t/\lambda_1))/(t/\lambda_1) - exp(-t/\lambda_1))$$

The yield curve according to this model has a long-term rate component (β_1), short-term component (2nd factor), and a 'hump' (3rd factor). The problem is to estimate the β_i and λ_1, for the curve which best fits the given market data: this is termed 'fitting the curve' to the data. Usually some form of sum of squared differences between the estimated curve and the market data is used as a measure of fit. The Nelson-Siegal-Svensson (NSS) model adds a further 'hump' term to the Nelson-Siegal model, with additional parameters β_4 and λ_2. This enables closer fitting of the curve to data in some cases, but also increases the computational cost of the fitting procedure.

The Nelson-Siegal model can be specified in UML by the following functions of *NelsonSiegalYC*:

```
static query nelsonsiegal(t : double,
          v1 : double, v2 : double, v3 : double,
          lambda1 : double) : double
```

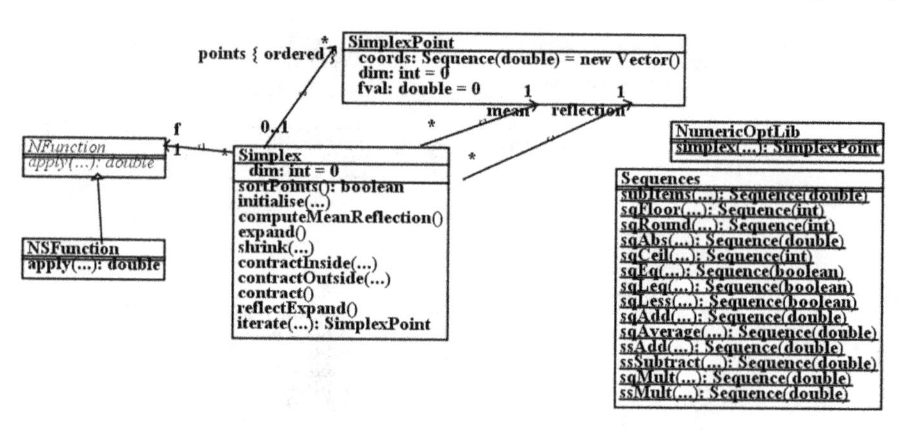

Fig. 7.5 Simplex class diagram

```
pre: t > 0 & lambda1 > 0
post:
  tscaled1 = t/lambda1 &
  exptscaled1 = (-tscaled1)->exp() &
  expratio1 = (1 - exptscaled1)/tscaled1 &
  result = v1 + v2*expratio1 +
           v3*(expratio1 - exptscaled1)
```

```
static query ns(t : double, v1 : double,
                v2 : double, v3 : double,
                lambda1 : double) : double
pre: t >= 0 & lambda1 > 0
post:
  (t = 0  =>  result = v1 + v2) &
  (t > 0  =>
     result = NelsonSiegalYC.nelsonsiegal(t,v1,v2,v3,lambda1))
```

The steps of the re-engineered estimation procedure are:

1. Create *InterestRate* instances for the market data
2. Produce a number of *NelsonSiegalYC* instances which provide possible estimated parameters to fit the data, using a commercially-confidential algorithm
3. Select the best of these, using a sum of squared differences measure of fit, and apply a numerical optimiser to refine the estimates.

For the second stage possible techniques include evolutionary algorithms such as genetic algorithms or differential evolution [3]. For the final stage we could use algorithms such as Matlab's `fminsearch` based on the simplex algorithm (Fig. 7.5).

The Nelder–Mead simplex algorithm performs a deterministic search to find a point where a minimal value of the associated n-ary function f is attained, $n > 1$. The attribute *dim* is the dimension n of the search space, the simplex consists of $n{+}1$ points, each point has n double-valued coordinates.

We can formalise the simplex algorithm in UML by the following operations of the Simplex class:

```
initialise(x : SimplexPoint)
post:
  x : points & dim = x.dim &
  Integer.subrange(1,x.dim)->forAll( i |
      SimplexPoint->exists( p |
          p.coords = ( x.coords.subrange(1,i - 1) ^
                     Sequence{ (x.coords->at(i)) * 1.05 } ^
                     x.coords.subrange(i + 1,x.dim) ) &
          p.dim = x.dim & p : points ) ) &
  points->forAll( p | p.fval = f.apply(p))
```

initialise generates an initial simplex including *dim* points based on 5% displacements in each dimension $i : 1..dim$ of the starting point x, and x is also included in the simplex points (so there are $dim + 1$ points in total). The f value of each point is then computed and stored in its *fval* attribute.

sortPoints sorts the points in ascending order of their f values:

```
sortPoints() : boolean
post:
  points = points@pre->sortedBy(fval) &
  result = true
```

The expression s defined by $col \rightarrow sortedBy(e)$ is a sequence formed by re-ordering collection *col* such that the elements of s are in non-descending order of the expression e evaluated on them, ie:

$$s[i].e \leq s[i+1].e$$

for $1 \leq i < col.size$. Usually e is an attribute of the objects in *col*, as in the above situation.

Hence the *points* sequence is ordered in terms of the degree of approximation to the point of minimum f value, with the best point (with lowest f value) being $points[1]$, and the worst being $points[dim + 1]$.

```
computeMeanReflection()
post:
  SimplexPoint->exists( m |
      m.coords = Sequences.sqAverage(points.front->collect(coords)) &
      m.dim = dim &
      m.fval = f.apply(m) &
      mean = m ) &
  SimplexPoint->exists( r |
      r.coords = Sequences.ssSubtract(
           Sequences.sqMult(mean.coords, 2.0), points.last.coords) &
      r.dim = dim &
      r.fval = f.apply(r) &
      reflection = r)
```

computeMeanReflection generates an average point *mean* of the *points.front* sequence (i.e., ignoring the worst point) and a reflection point:

$$reflection = 2 * mean - points[dim + 1]$$

Vector computations on points are performed by vector algebra upon their coordinate sequences, using operations *ssAdd*, *ssSubtract*, etc, from the *Sequences* library.

The main part of the algorithm then uses various strategies to replace the worst point $points[dim + 1]$ with an improved point with lower f value, or in the final case (*shrink*) to contract the simplex around the best point. We always store $f(p)$ in *p.fval* for a simplex point p when p is created or updated, in order to avoid recomputation of f on p: this is another form of function caching.

```
reflectExpand()
post:
  (points.first.fval <= reflection.fval &
   reflection.fval < points[dim].fval  =>
         points[(dim+1)] = reflection) &
  (reflection.fval < points.first.fval  =>
         self.expand()) &
  (points[dim].fval <= reflection.fval  =>
         self.contract())

expand()
post:
  SimplexPoint->exists( s |
     s.coords = Sequences.ssAdd(
        mean.coords, Sequences.sqMult(
           Sequences.ssSubtract(mean.coords,
                 points.last.coords), 2.0)) &
     s.dim = dim &
     s.fval = f.apply(s) &
     ((s.fval < reflection.fval => points[(dim+1)] = s) &
      (reflection.fval <= s.fval => points[(dim+1)] = reflection)))

contractOutside(c : SimplexPoint)
post:
  (c.fval < reflection.fval => points[(dim+1)] = c) &
  (reflection.fval <= c.fval => self.shrink(points.first))

contractInside(cc : SimplexPoint)
post:
  (cc.fval < points.last.fval => points[(dim+1)] = cc) &
   (points.last.fval <= cc.fval => self.shrink(points.first))

contract()
post:
  (reflection.fval < points.last.fval  =>
     SimplexPoint->exists( c |
         c.coords = Sequences.ssAdd(mean.coords,
             Sequences.sqMult(
                 Sequences.ssSubtract(reflection.coords,
                          mean.coords), 0.5)) &
```

```
          c.dim = dim &
          c.fval = f.apply(c) &
          self.contractOutside(c)) ) &
  (reflection.fval >= points.last.fval  =>
       SimplexPoint->exists( cc |
           cc.coords = Sequences.ssAdd(mean.coords,
                Sequences.sqMult(
                   Sequences.ssSubtract(points.last.coords,
                          mean.coords), 0.5)) &
            cc.dim = dim &
            cc.fval = f.apply(cc) &
            self.contractInside(cc) ) )

shrink(x1 : SimplexPoint)
post:
  Integer.subrange(2, dim+1)->forAll( i |
      points[i].coords =
         Sequences.ssAdd(x1.coords,
            Sequences.sqMult(
                Sequences.ssSubtract(points[i].coords,
                         x1.coords), 0.5) ) &
      points[i].fval = f.apply(points[i]) )


iterate(tol : double) : SimplexPoint
activity:
  execute sortPoints() ;
  while diameter() > tol
  do
    execute (
      computeMeanReflection() &
      reflectExpand() &
      sortPoints()  )  ;
  return points[1]
```

diameter() measures the diameter of the simplex, the distance between *points*[1] and *points*[*dim* + 1].

The Matlab *fminsearch*(f, x) call for a double-valued function f of two or more double-valued parameters is expressed as the invocation *NumericOptLib.simplex* (f, x, 0.0001) of the simplex algorithm:

```
static simplex(f : NFunction, x : Sequence(double),
               tol : double) : SimplexPoint
pre: tol > 0
post:
  SimplexPoint->exists( p |
     p.dim = x.size &
     p.coords = x &
     Simplex->exists( sx |
         sx.f = f &
         sx.dim = x.size &
         sx.initialise(p) &
         result = sx.iterate(tol) ) )
```

For NS estimation, the definition of *apply* in NSFunction is the sum of squared differences of the NS curve from the market data points (*r.maturity*, *r.rate*):

```
query apply(p : SimplexPoint) : double
post:
  beta1 = p.coords[1] &
  beta2 = p.coords[2] &
  beta3 = p.coords[3] &
  lambda = p.coords[4] &
  result = InterestRate->collect( r |
      ( r.rate - NelsonSiegalYC.ns(r.maturity, beta1, beta2,
                          beta3, lambda) )->sqr() )->sum()
```

Instead of using yields, market price data can be used to fit the curve. In this case, the simplex could operate with a function object with *apply* defined as the sum of squared differences between the actual bond prices and those obtained from the curve.

We tested the above procedure with an example of eight zero-coupon bonds with durations ranging from 1 to 10 years. Using yields to measure the degree of fit, we obtained the following initial parameter values for the market data given above:

$$\begin{aligned}\beta_1 &= 0.0619589381 0393646\\ \beta_2 &= -0.061204959672191465\\ \beta_3 &= 0.1289089344485046\\ \lambda_1 &= 2.0\end{aligned}$$

The accuracy (sum of squared differences to the market data yields) of this estimated curve is $1.8 * 10^{-4}$.

After refinement using the simplex algorithm, the parameters become:

$$\begin{aligned}\beta_1 &= 0.06594219551067752\\ \beta_2 &= -0.06593501071223262\\ \beta_3 &= 0.12226954660991908\\ \lambda_1 &= 1.8624443350354993\end{aligned}$$

The accuracy is improved to $6 * 10^{-6}$.

The curve produced from these parameters is shown in Fig. 7.6.

A similar procedure can be defined for NSS curve estimation, using simplexes with 6 points instead of 4.

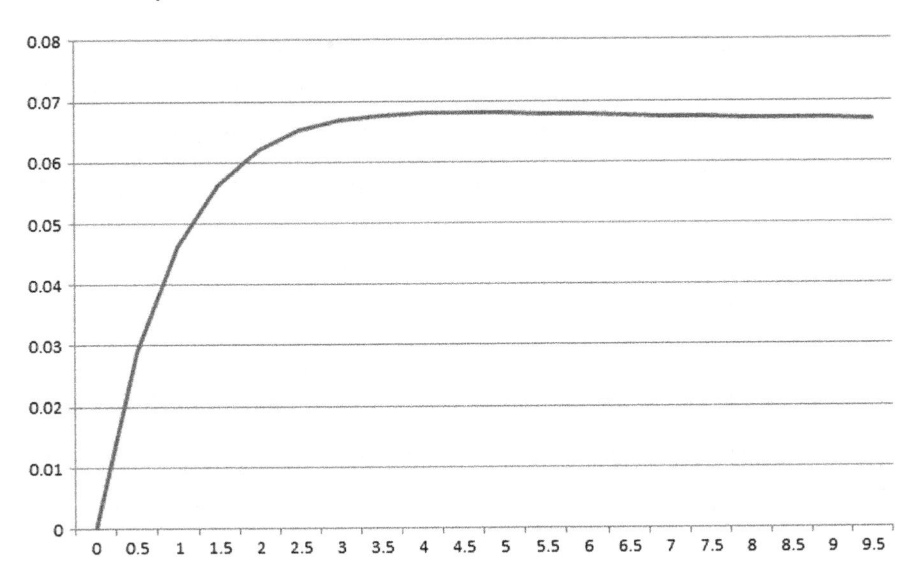

Fig. 7.6 Estimated yield curve

Summary

In this chapter we have described how MBD and UML can be used to support the modernisation of financial software via re-engineering. The process has been illustrated with a practical example of a Matlab to C# re-engineering project.

Exercises

1. Evaluate $y(n * \lambda_1)$ for the Nelson-Siegal model, for positive integer n.
   Since $y(t) \geq 0$ is normally expected for any $t \geq 0$, what constraints can be inferred relating the β_i?

2. Let the $slope(n : int) : double$ of a NS curve be defined as $(y(n * \lambda_1) - y(0))/n$.
   Express the definition of *slope* as an operation of *NelsonSiegalYC*.

3. What are the approximate conditions for $slope(n)$ to be (i) positive; (ii) negative, when $n > 10$?

4. We have discussed the spot interest rate curve in this chapter. Also of concern in finance is the *forward interest rate* curve.
   The forward interest rate $fr(n)$ is the annual interest rate for a year n years in the future, deduced from the spot rates $y(n)$ and $y(n+1)$.
   Given an investment of amount N held for duration $n+1$, N should have gained in value to $N * e^{(n+1)*y(n+1)}$ at its term. But this can also be divided into an investment

of N for duration n at rate $y(n)$, and a 1-year investment of $N * e^{n*y(n)}$ at rate $fr(n)$. Derive an equation for $fr(n)$ in terms of y and n, assuming continuous compounding.

5. Assume a NS model for $y(t)$ and express $fr(t)$ as a formula in the NS terms.

6. What are the efficiency trade-offs in using bond prices versus using yields to fit a yield curve?

References

1. S. Yassipour-Tehrani, K. Lano, S. Zschaler, *Requirements Engineering in Model Transformation Development: An Interview-Based Study* (ICMT, 2016)
2. M.B. Nakicenovic, An Agile Driven Architecture Modernization to a Model-Driven Development Solution. Int. J. Adv. Softw. **5**(3, 4) 308–322 (2012)
3. M. Gilli, S. Grosse, E. Schumann, *Calibrating the Nelson-Siegal-Svensson Model*, COMISEF Working Paper WPS-031 (2010)

Chapter 8
Agile Model-Based Development Approaches

In this chapter we describe:

- Existing agile MBD approaches
- Guidelines for introducing agile MBD
- How to define and use DSLs for finance.

8.1 Existing Agile MBD Approaches

A number of development approaches have been defined and used to combine the benefits of agile and MBD approaches [1]:

- MDD-SLAP: used in the telecomms domain (by Motorola). UML and SDL models are used (sequence diagrams, state machines, class diagrams). Each iteration is divided into three stages: requirements/architecture, development, integration/testing [2].
- SunGard approach: used in the finance domain (trading systems). The approach uses XML models, and the Scrum and Kanban methods [3].
- Hybrid MDD: used for web applications. The approach uses XML models, and XSLT transformations [4].
- Volvo Car Group: used for automotive systems, with Simulink and device models for vehicle ECUs, which are refined in short iterations [5].
- UML-RSDS: used for transformations and financial systems. It uses UML models (primarily class diagrams, use cases and OCL), it follows a Scrum process [6] with three-phase iterations.

Generally, benefits have been found from agile MBD in providing improved stakeholder collaboration and improved reusability [3], and improved flexibility in development [5]. The combination has also improved the usability of MBD within contexts where there is not established expertise in MBD [3].

K. Lano and H. Haughton, *Financial Software Engineering*, Undergraduate Topics in Computer Science, https://doi.org/10.1007/978-3-030-14050-2_8

8.2 Guidelines for Introducing Agile MBD

The introduction of MBD can be difficult due to the complexity of general-purpose modelling languages such as UML, which consequently have complex supporting tools and high training costs. A focus on agility and lightweight modelling can make the use of MBD more feasible, by either (i) using only an essential subset of UML oriented to a particular application area/platform; (ii) using XML or a similar minimal modelling language; (iii) using a domain-specific language (DSL), i.e., a modelling language constructed for the particular application area.

UML-RSDS represents approach (i). This has the advantage of using a fully-featured industry-standard modelling language, and avoids the effort needed to create a DSL. Approach (ii) suffers from the limitations of XML, which is oriented to text representation of tree-structured data. Regarding (iii), industry surveys have found successful introduction of MBD in cases where modelling is used to define DSLs and code templates (model to text transformations) for narrow domains. The DSL and templates encode expertise in the domain. If organisations are able to take a long-term view, the DSL approach enables them to use MBD for product lines. Libraries of functions and components for the domain can be established and maintained to form a domain *platform*, upon which new applications can be built. For example, operations $\mathit{calculateMaturityPrice}(\mathit{now} : \mathit{double}, r : \mathit{double}) : \mathit{double}$ and $\mathit{calculateValue}(\mathit{now} : \mathit{double}, r : \mathit{double}) : \mathit{double}$ based on Black–Scholes option pricing can be added to the domain model for financial assets and derivative securities to form a platform for this domain (Fig. 8.1).

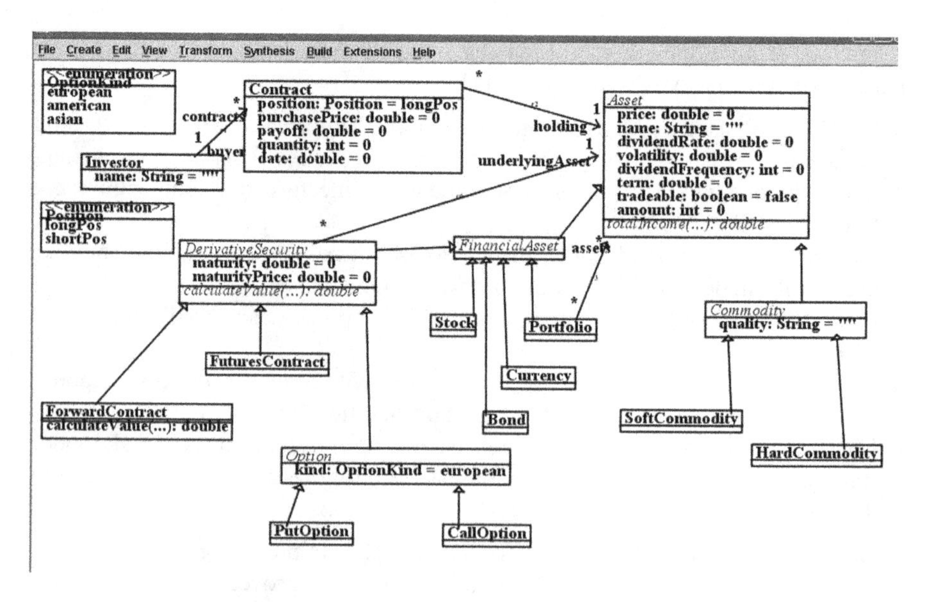

Fig. 8.1 Derivative securities DSL

Useful practices for agile MBD include:

- Modelling as coding, with executable models serving both as specifications and as implementations.
- Performing refactoring, quality analysis and testing at the specification level.
- Using text-based model editing tools to accelerate development compared to graphical editors.
- Paired modelling (e.g., a developer and a customer representative both work together on the same model).
- A parallel tooling team supporting the main development team. This separates the MBD tooling work from the main application development activity and means that the main team does not need to have the specialised expertise required for MBD tool development.
- Extending the scope of model-based construction of elements, avoiding manual coding wherever possible.
- Reusing existing systems and libraries wherever possible.
- Using MBD experts to train and lead teams.

Key issues include the high learning curve for developers (for MBD), and the need for traceability between models and code. Integrating/merging changes in multiple models is a potential problem with the use of MBD and agile MBD—this can be reduced by minimising the number of independent models. It is necessary to establish an effective tool chain of interoperable MBD tools to support a project. We have found graphical editing of models too slow in practice, and hence we recommend the use of a text-based editor to support rapid revision of large application models. Automatic layout tools can be used to produce visual representations of the models when needed.

Note that while UML-RSDS includes a substantial subset of UML 2, it does not include mechanisms for real time specification, such as setting execution time bounds on operations. It does not include mechanisms for specification of explicit parallelism.

Existing practice in financial software development involves prototyping in a language such as Excel, Matlab or Python, and then either (i) retaining these prototypes as production software, or (ii) using manual coding to produce versions in a programming language such as C++ or C#. Option (i) has deficiencies in terms of software quality and efficiency, whilst option (ii) is costly, and involves duplicating the software functionality. Our approach retains the specification and prototype, but uses automated code generation to produce an efficient production implementation from these.

8.3 Defining and Using DSLs

DSLs are used to express *domains*, including business domains (e.g., financial derivatives) or technical domains (e.g., EIS architectures). DSLs capture the common elements and key concepts of the domain and organise these into a language which can be reused as the basis for the specification of many different applications within the domain.

Transformations can read DSL models and produce code or other artifacts. The transformations automate previously manual development processes and steps. Transformations together with the DSL encode domain knowledge and expertise. DSLs also enable application modelling using terminology familiar to stakeholders.

To construct a DSL the following steps are typically taken:

- Identify the key terminology/concepts in a specific domain (the domain ontology).
- Model these concepts in a class diagram—this defines the abstract syntax of the DSL as a metamodel/language. Operations and use cases can be included, if they represent functionality and procedures which are significant in the domain (and which are not already defined in a library).
- Identify a concrete syntax for DSL models, which is convenient for stakeholders and for automated processing. A text format is also needed for the *serialisation syntax* used to store/transmit models.
- Define model parsers and transformations operating on DSL models, to produce other required artifacts, such as code, test cases, model analyses or graphical representations.

Using UML-RSDS, DSLs can be defined as class diagrams. A default concrete syntax is then provided by the textual form of instance models of these class diagrams. Such models are stored in text files consisting of lines of the form (i)

```
obj : E
```

expressing that *obj* is an instance of DSL entity E, or (ii)

```
obj.f = value
```

expressing that the value of the single-valued feature f of *obj* is *value*, or (iii)

```
obj : x.f
```

expressing that *obj* is an element of the collection-valued feature f of instance x. This text format is also the default serialisation syntax (an XML format is also available). UML-RSDS transformations can read and write such model files, and therefore can perform tasks such as code generation or semantic mapping of the models.

8.3.1 DSL Example: Derivative Securities

We can construct a DSL based on the standard financial terminology and concepts in the domain of derivative securities (Fig. 8.1):

- The entities include *Derivative security*, with specialisations *Option*, *Forward contract*, *Futures contract*, etc.
- Other entities could include: *Trader*, *Contract*, *Investor*, *Exchange*, *Asset*, *Commodity*, *Stock*, *Margin account*, etc.
- Associations and attributes could include: *underlyingAsset*, *repo rate*, *value*, *futures price*, *expiration date*, *delivery price*, *strike price*, *spot price*, etc.
- Operations can be defined to compute the value and maturity price of each category of derivative security.

Ideally, the most experienced available domain experts should advise on the DSL model elements and structure. After defining the class diagram, it should be reviewed and refined based on its intended uses. Refactoring can be used to improve the quality of the DSL, to remove redundancies and simplify its structure.

The *Contract* class expresses properties intrinsic to a particular contract (such as its position) and properties which may typically be open to negotiation between an investor and the supplier of a financial product (asset). For example, the purchase price and date. In contrast, *Asset* and its subclasses contain properties that are intrinsic to the product and cannot be negotiated, such as the volatility. We have rationalised the diagram to avoid duplication, for example the attribute *maturityPrice* of *DerivativeSecurity* expresses the common concept of delivery, futures and strike prices found in specific kinds of derivative security. These separate attributes were merged as a result of analysis and refactoring.

A concrete representation for derivative securities could be spreadsheet data tables (Table 8.1).

Other model data (Investors, Contracts, Assets) can also be represented in a similar way. CSV-format spreadsheet models can be easily parsed and mapped to instance models in UML-RSDS. An alternative would be the standard object representation notation:

```
a : Stock
a.name = "IBM"
a.amount = 100
a.tradeable = true
a.price = 145.16
```

Table 8.1 DSL concrete syntax

Id	Type	Maturity	Maturity price	Underlying asset Id
"1"	"calloption"	2021.0	125	"3"

```
x : CallOption
x.amount = 1
x.maturity = 2021.0
x.maturityPrice = 125
x.underlyingAsset = a
```

The model can represent common situations such as hedging using derivatives. For example, if an investor in shares wants to hedge the risk of the unit price of a main contract $c : Contract$ holding an equity asset a with quantity $c.quantity$ being below a level P at a future date $now + t$, they can also enter into a supplementary contract $cf : Contract$ to hold American put options f with $f.maturity = now + t$, $cf.quantity = c.quantity, f.underlyingAsset = a$, and $f.maturityPrice$ equal to P. At the future date, if the share price is $Q > P$, then the investor will not exercise the options, and instead can sell quantity $c.quantity$ of a at Q per unit. Their profit is only reduced by the cost V of the options. On the other hand, if $Q < P$, they can exercise the options and sell quantity $cf.quantity$ at P (because of contract cf), making a profit of $(P - Q) * c.quantity - V$.

The above assumes that one futures contract is for one share, in practice a contract for a derivative such as an option based on a share would be an option to sell or purchase multiple (typically 100) shares.

The value of each security can be calculated using a $calculateValue(now, r)$ operation, where now represents the current date, and r is the relevant risk-free interest rate. The operation is defined for each kind of derivative security. For long forward contracts, we have the general equation (using continuous compounding):

$$value = (S - I) * e^{-q*dt} - K * e^{-r*dt}$$

where S is the current price of the underlying asset, q is the continuous dividend yield rate of the asset, T is the maturity date (years), t the current date, $dt = T - t$, and I is the present value of the income from fixed payments from the asset over the period dt (for example, coupon payments in the case of a coupon bond), K is the maturity price, and r is the relevant risk-free rate of interest. The reason for this equation is that the forward contract plus $K * e^{-r*dt}$ in cash (invested at rate r) is equivalent as an investment to buying a proportion e^{-q*dt} of the share now (for $S * e^{-q*dt}$) and borrowing amount $I * e^{-q*dt}$. In the first case, by maturity, K has been earned from the cash and the share can be bought with this. Conversely, if the share fraction had been bought and all dividend income re-invested in purchasing more of the share, the holding would have grown to one share. The fixed payments from the share fraction would also have paid off the borrowed fraction of I. Hence the value of this investment at maturity is also one share.

Therefore, we can specify the operation definition:

```
ForwardContract::
query calculateValue(now : double, r : double) : double
post:
  dt = maturity - now &
```

```
maturityValue = maturityPrice*((-r*dt)->exp()) &
income = underlyingAsset.totalIncome(now, maturity) &
dividendYield = underlyingAsset.dividendRate &
result = (underlyingAsset.price - income) *
                  ((-dividendYield*dt)->exp()) - maturityValue
```

This definition also applies for *FuturesContract*. For short forward/future contracts the value is $-calculateValue(now, r)$.

The maturity price K which makes the present value 0 is given by:

$$K = (S - I) * e^{(r-q)*dt}$$

and this can be expressed as an operation

```
ForwardContract::
query calculateMaturityPrice(now : double, r : double) : double
post:
  income = underlyingAsset.totalIncome(now, maturity) &
  dividendYield = underlyingAsset.dividendRate &
  result = (underlyingAsset.price - income)*
            ((r-dividendYield)*(maturity-now))->exp()
```

This is valid for both long and short positions in the forward/future contract.

For options, we can reuse library operations such as

$$FinLib.euPutOptionPrice(underlyingAsset.price, maturityPrice, r, \\ underlyingAsset.dividendYield, maturity - now, \\ underlyingAsset.volatility, income)$$

to value the security. These are defined using the solutions for the Black–Scholes equation [7].

According to the Black–Scholes analysis, a European call option with term $dt = T - t$, on a non-divided paying share asset with price S has the current value

$$c = S * N(d_1) - X * e^{-r*dt} * N(d_2)$$

where N is the cumulative probability distribution function for the normal distribution with mean 0 and standard deviation 1. d_1 and d_2 are defined in terms of S, X (the maturity price), $dt = T - t$, r and σ, the volatility of the underlying asset. This is also the value of an American call option (under the assumption that it is never optimal to exercise such an option prior to maturity).

$$d_1(S, X, dt, r, \sigma) = ((S/X) \rightarrow log() + (r + \sigma.sqr/2) * dt)/(\sigma * dt.sqrt)$$

and

$$d_2(S, X, dt, r, \sigma) = d_1(S, X, dt, r, \sigma) - \sigma * dt.sqrt$$

A European put option on such an asset has the value

$$p = X * e^{-r*dt} * N(-d_2) - S * N(-d_1)$$

There is no precise formula for an American put option, but various approximation techniques can be used to compute it.

If the asset pays an known income over its lifetime (e.g., a share that pays known cash dividends at certain time points), then the present value I of the income is subtracted from S:

$$c = (S - I) * N(d_1') - X * e^{-r*dt} * N(d_2')$$

where d_1' and d_2' are computed as $d_1(S - I, X, dt, r, \sigma)$ and $d_2(S - I, X, dt, r, \sigma)$.

The corresponding put valuation is

$$p = X * e^{-r*dt} * N(-d_2') - (S - I) * N(-d_1')$$

If instead of a specific income, the asset pays a continuous dividend (*dividendRate* in Fig. 8.1) equal to q, we have the valuations:

$$c = S * e^{-q*dt} * N(d_1'') - X * e^{-r*dt} * N(d_2'')$$

and

$$p = X * e^{-r*dt} * N(-d_2'') - S * e^{-q*dt} * N(-d_1'')$$

where d_1'' and d_2'' are computed as $d_1(S, X, dt, r - q, \sigma)$ and $d_2(S, X, dt, r - q, \sigma)$.

The same formula applies to assets which are foreign currencies, with q being the risk-free interest rate in the foreign currency.

A simple approach to bond options is to value them in a similar way to share options: in the case of a zero-coupon bond, S is taken as the bond price, σ as the volatility of this price, and r as the risk-free interest rate for an investment of duration dt. For a coupon bond, the income I from the bond is evaluated as the discounted sum of the coupon payments during the interval dt and then the formulae applied with $S - I$ instead of S.

euPutOptionPrice can therefore be defined as:

```
static query euPutOptionPrice(s : double, x : double,
    r : double, q : double, dt : double,
    sigma : double, income : double) : double
pre: sigma > 0 & dt > 0
post:
  adjustedS = (s - income)*((-q*dt)->exp()) &
  d1 = FinLib.bsd1(s-income,x,dt,r-q,sigma) &
  d2 = FinLib.bsd2(s-income,x,dt,r-q,sigma) &
  result =
    x*((-r*dt)->exp())*NormalDist.cumulative(-d2) -
```

```
                adjustedS*NormalDist.cumulative(-d1)
```

Where:

```
static query bsd1(s : double, x : double, dt : double,
      r : double, sigma : double) : double
pre: sigma > 0 & dt > 0
post:
  result = ((s/x)->log() +
            (r + sigma.sqr/2.0)*dt)/(sigma*dt.sqrt)
```

and

```
static query bsd2(s : double, x : double, dt : double,
      r : double, sigma : double) : double
pre: sigma > 0 & dt > 0
post:
  result = FinLib.bsd1(s,x,dt,r,sigma) - sigma*dt.sqrt
```

Note that if a called operation has a precondition P, any caller of the operation has to ensure P at the point of call, which in this case is done by repeating the precondition P as the caller's precondition.

Summary

In this chapter, we have reviewed existing agile MBD approaches and identified guidelines for introducing agile MBD in practice. We also illustrated agile MBD techniques and an example of using DSLs with agile MBD.

Exercises

1. Considering the definition of *value* for long forward contracts given above, how does this value change as the difference *dt* decreases?

2. Put-call parity is the relationship

$$c + X * e^{-r*dt} = p + S$$

for the values p and c of European put and call options on the same stock (for non-dividend paying shares). Use this relationship to give a definition for *euCallOptionPrice* in terms of *euPutOptionPrice*.

3. A *bull spread* is a combination of two call options on the same amount of the same stock with the same maturity date. This consists of a long position in the first option, and a short position in the second, with the first option having a lower strike price than the second. Express a bull spread in terms of the derivative securities DSL.

References

1. S. Hansson, Y. Zhao, H. Burden, How MAD are we?: empirical evidence for model-driven agile development (2014)
2. Y. Zhang, S. Patel, Agile model-driven development in practice. IEEE Softw. **28**(2), 84–91 (2011)
3. M.B. Nakicenovic, An agile driven architecture modernization to a model-driven development solution. Int. J. Adv. Softw. **5**(3, 4), 308–322 (2012)
4. G. Guta, W. Schreiner, D. Draheim, A lightweight MDSD process applied in small projects, in *Proceedings 35th Euromicro Conference on Software Engineering and Advanced Applications* (IEEE, 2009)
5. U. Eliasson, R. Heldal, J. Lantz, C. Berger, *Agile MDE in Mechatronic Systems – An Industrial Case Study, MODELS* (2014)
6. K. Schwaber, M. Beedble, *Agile Software Development with Scrum* (Pearson, London, 2012)
7. F. Black, M. Scholes, The pricing of options and corporate liabilities. J. Polit. Econ. **81**, 637–659 (1973)

Chapter 9
Analysis of Financial Products: CDOs

In this chapter we give a detailed example of mathematical analysis of a particular financial product—collateralised debt obligations (CDOs)—and show how the theoretical results obtained can be expressed as software specifications.

We revisit the seminal work of Davis and Lo [1] on CDO analysis, and expand on related work by Hammarlid [2] in regards to a moment generator (based on assuming a Poisson number of outbreaks) to determine risk contributions to a CDO.

9.1 Introduction

In this chapter we expand on the proofs provided by Davis and Lo in [1] regarding Theorem 1 of their paper. Although no new results are obtained (as it relates to [1]), we develop the theorem to illustrate the way that such results can be obtained. We also determine sector risk contributions in the framework of the Hammarlid model [2] and derive a simple formula for these contributions.

9.2 Preliminaries

In [1] the authors consider a portfolio of n identical bond/loan securities (not necessarily independent of each other). Each such security is assumed to have the same default probability, recovery rate etc. A random variable Z_i, $i = 1..., n$ taking the value $Z_i = 1$ if bond i defaults and $Z_i = 0$ otherwise is used to model the default status of a single borrower i. Later, in this section, it is shown that such defaults could be caused either by a direct default of borrower i or by contagion due to the default of another borrower $j \neq i$.

K. Lano and H. Haughton, *Financial Software Engineering*, Undergraduate Topics in Computer Science, https://doi.org/10.1007/978-3-030-14050-2_9

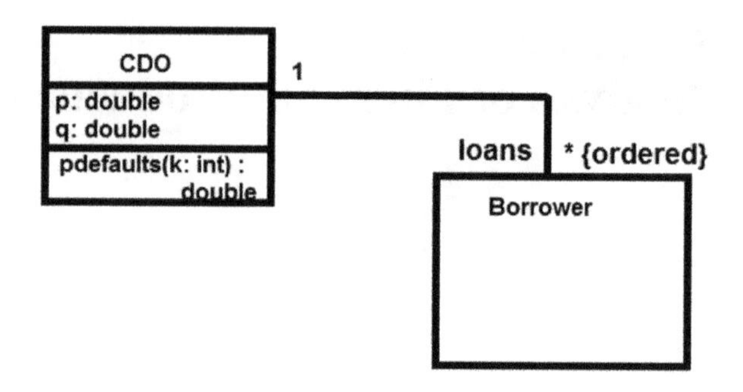

Fig. 9.1 Simple CDO model

With the above setup, the total number of borrowers defaulting is thus given by

$$N = Z_1 + Z_2 + \cdots + Z_n. \tag{9.1}$$

Let X_i, Y_{ji} be independent Bernoulli random variables denoting the default status of a borrower i (i.e., $X_i = 1$ if i defaults or $X_i = 0$ otherwise) and whether default of borrower j affects borrower i (i.e., $Y_{ji} = 1$) or otherwise (i.e., $Y_{ji} = 0$) with

$$P[X_i = 1] = p$$
$$P\left[Y_{ji} = 1\right] = q.$$

This corresponds to a model of CDOs in which the probability of borrower defaults is the same, p, for every borrower, so this probability can be defined as an attribute in the *CDO* class (Fig. 9.1), and likewise for the probability of infection, q. Loan i is represented by *loans*[i] in Fig. 9.1, and n is *loans.size*.

The random variable Z_i can now be defined:

$$Z_i = X_i + (1 - X_i)\left(1 - \Pi_{j \neq i}\left(1 - X_j Y_{ji}\right)\right). \tag{9.2}$$

From Eq. 9.2 we note that if $X_i = 1$ (i.e., i defaults) then it does not matter whether any other borrower j defaulting affects i (i.e., $X_j = 1$ and $Y_{ji} = 1$) since the first bracketed term on the right side of the equation guarantees that only a direct default contributes to $Z_i = 1$. Conversely if $X_i = 0$ (i.e., i does not directly default) then in order for $Z_i = 1$ it is required that $X_j = 1$ and $Y_{ji} = 1$ for at least one $j \neq i$.

As the joint distribution of $(Z_1, \ldots, Z_n)$ has the same joint distribution as $\left(Z_{\sigma(1)}, \ldots, Z_{\sigma(n)}\right)$ for any finite permutation σ of the indices 1, 2, 3... we say that $Z_i, i = 1, \ldots, n$ is exchangeable.

9.3 Davis and Lo Theorem

In [1] Davis and Lo proved the following result.

Theorem 9.3.1 *The distribution function F of the probability P[N=k] of k defaults is defined by*

$$F(n,k,p,q) = C_k^n \alpha_{nk}^{pq} = P[N=k], \tag{9.3}$$

where

$$\alpha_{nk}^{pq} = p^k (1-p)^{n-k} (1-q)^{k(n-k)} + \sum_{i=1}^{k-1} C_i^k p^i (1-p)^{n-i} \left(1-(1-q)^i\right)^{k-i} (1-q)^{i(n-k)}.$$

Proof

Note that since the Z_i, $i = 1, \ldots, n$ are exchangeable we can write the event indicating that k borrowers have defaulted as $Z_1 = 1, \ldots, Z_k = 1, Z_{k+1} = 0, \ldots, Z_n = 0$ since all other permutations of the *Z's* will produce the same probabilities. This event can be achieved in one of 2 ways: either the k borrowers default directly (i.e., $X_i = 1, i = 1, \ldots, k$) and do not cause any contagion with the other bonds (i.e., $Y_{ji} = 0, i = 1, \ldots, k, j = k+1, \ldots, n$) or a subset of the k borrowers (i say, where $i < k$) directly default and spread contagion to the remaining $(k-i)$ borrowers but not to the borrowers corresponding to indices $k+1, \ldots, n$.

For the first case, since we assume that k borrowers have defaulted directly (each with probability p) there must be $n-k$ that do not default directly (each with probability $1-p$). The joint probability of these two events is $p^k (1-p)^{n-k}$. As the k defaults do not affect the other borrowers this implies that there are $k(n-k)$ combinations where there is no contagion (with probability $1-q$) resulting in the probability $(1-q)^{k(n-k)}$. Since the *X's* and *Y's* are independent the joint probability is just the product:

$$p^k (1-p)^{n-k} (1-q)^{k(n-k)}.$$

For the second case, we assume that given an i where $i < k$ borrowers directly default (each with probability p) and hence $n-i$ do not default directly. The probability that this event occurs is $p^i (1-p)^{n-i}$. Note that since the i direct defaulters only affect $k-i$ of the first borrowers then there will be $n-k$ borrowers not affected by contagion (each with probability $1-q$). This implies that there are $i(n-k)$ combinations where there is no contagion with probability $(1-q)^{i(n-k)}$. Given that i defaulters affect $k-i$ via contagion we should account for the i that are not affected by contagion (which are those that default directly) which is given by the probability $(1-q)^i$. Hence given this probability, the probability that a borrower defaults via way of contagion is given by $1-(1-q)^i$. Since there are $k-i$ of these affected borrowers the total probability is $\left(1-(1-q)^i\right)^{k-i}$. Combining the probabilities we get $p^i (1-p)^{n-i} \left(1-(1-q)^i\right)^{k-i} (1-q)^{i(n-k)}$.

Note that in assuming that there are i borrowers defaulting (this i is arbitrary) we should account for the fact that the above analysis is true for any $i < k$, i.e., $i = 1, \ldots, k - 1$. Further, since there are C_i^k ways of choosing the i from the k and given exchangeability the total probability of the second case is

$$\sum\nolimits_{i=1}^{k-1} C_i^k p^i (1-p)^{n-i} \left(1-(1-q)^i\right)^{k-i} (1-q)^{i(n-k)} .$$

Finally we note that there are C_k^n ways of choosing the fixed total number of defaulters k from the n and hence the total probability $P[N = k]$ is as given in the theorem. □

This result can be used to define the probability of k defaults in the model of Fig. 9.1 by the operation

```
CDO::
query pdefaults(k : int) : double
pre: k >= 0
post:
  n = loans.size &
  result = MathLib.combinatorial(n,k)*alpha(n,k)
```

Where:

```
CDO::
query alpha(n : int, k : int) : double
pre: k >= 0
post:
  directDefault =
    (p->pow(k))*((1-p)->pow(n-k))*((1-q)->pow(k*(n-k))) &
  indirectDefault = Integer.Sum(1,k-1,i,
             MathLib.combinatorial(k,i) *
             (p->pow(i)) * ((1-p)->pow(n-i)) *
             ((1 - (1-q)->pow(i))->pow(k-i)) *
             ((1-q)->pow(i*(n-k)))) &
  result = directDefault + indirectDefault
```

This definition is a direct transcription of the theorem. To avoid excessive complexity in the specification, the *alpha* function can be further factored by defining an auxiliary function

```
CDO::
query defaultInfection(n : int, k : int, i : int) : double
pre: i > 0 & k >= 0
post:
  result = MathLib.combinatorial(k,i) *
      (p->pow(i))*((1-p)->pow(n-i)) *
```

```
((1 - (1-q)->pow(i))->pow(k-i))*((1-q)->pow(i*(n-k)))
```

And calling this in the *Sum* operator in *alpha*:

```
CDO::
query alpha(n : int, k : int) : double
pre: k >= 0
post:
  directDefault =
    (p->pow(k))*((1-p)->pow(n-k))*((1-q)->pow(k*(n-k))) &
  indirectDefault =
    Integer.Sum(1,k-1,i,defaultInfection(n,k,i)) &
  result = directDefault + indirectDefault
```

Requirements validation includes checking that this computational version of the formula agrees with the mathematical formula. The specification can also be executed to check computed results.

9.4 Expectation and Variance

In [1] the expectation of the number of defaults $E[N]$ is shown to be

$$E(N) = n\left(1 - (1-p)(1-pq)^{n-1}\right).$$

Proof

Let $N = Z_1 + Z_2 + \cdots + Z_n$ and hence $E[N] = E[Z_1 + Z_2 + \cdots + Z_n]$.
Now,

$$Z_1 + Z_2 + \cdots + Z_n = X_1 + X_2 + \cdots + X_n \left(= \sum_{i=1}^{n} X_i\right)$$
$$+ \sum_{i=1}^{n} (1 - X_i)\left(1 - \Pi_{j \neq i}\left(1 - X_j Y_{ji}\right)\right).$$

Since the X's are independent, $E[X_1 + X_2 + \cdots + X_n] = E[X_1] + E[X_2] + \cdots E[X_n]$. Moreover, since each X_i denotes a Bernoulli random variable we have

$$E[X_i] = 1 \cdot p + 0 \cdot (1-p) = p$$

and hence $E\left[\sum_{i=1}^{n} X_i\right] = np$.

Since the X's and Y's are independent

$$E\left[\sum_{i=1}^{n}(1-X_i)\left(1-\Pi_{j\neq i}\left(1-X_jY_{ji}\right)\right)\right] = \sum_{i=1}^{n}E\left[(1-X_i)\left(1-\Pi_{j\neq i}\left(1-X_jY_{ji}\right)\right)\right].$$

Again, due to independence, the right hand side of the above equation is

$$\begin{aligned}\sum_{i=1}^{n}E\left[(1-X_i)\left(1-\Pi_{j\neq i}\left(1-X_jY_{ji}\right)\right)\right] &= \sum_{i=1}^{n}E\left[1-X_i\right]E\left[1-\Pi_{j\neq i}\left(1-X_jY_{ji}\right)\right] \\ &= \sum_{i=1}^{n}(1-p)\left(1-(1-pq)^{n-1}\right) \\ &= n\left((1-p)\left(1-(1-pq)^{n-1}\right)\right).\end{aligned}$$

The first product in the second equality above follows from the linearity of the expectation operator. The second product in the second equality follows from the fact that

$$E\left[1-X_jY_{ji}\right] = 1-pq$$

and since $j \neq i$ there are $n-1$ possibilities for j.

Putting together the above results we get

$$\begin{aligned}E[N] &= np + n\left((1-p)\left(1-(1-pq)^{n-1}\right)\right) \\ &= n\left(1-(1-p)(1-pq)^{n-1}\right).\end{aligned}$$

□

This gives us an operation

```
CDO::
query edefaults() : double
post:
  n = loans.size &
  result = n*(1-(1-p)*((1-p*q)->pow(n-1)))
```

For the model of Fig. 9.1.

Theorem 9.4.1 *The variance var* $[N]$ *in [1] is*

$$var[N] = E[N] + n(n-1)\beta_n^{pq} - (E[N])^2$$

where

$$\beta_n^{pq} = p^2 + 2p(1-p)\left(1-(1-q)(1-pq)^{n-2}\right) + (1-p)^2\left(1-2(1-pq)^{n-2}+\left(1-2pq+pq^2\right)^{n-2}\right).$$

Proof

The variance can be written as

$$var[N] = E\left[N^2\right] - (E[N])^2.$$

We have already calculated $E[N]$ so it remains to determine $E\left[N^2\right]$. Note that

$$E\left[N^2\right] = E[Z_1 + Z_2 + \cdots + Z_n]^2. \tag{9.4}$$

Note that the right hand side of Eq. 9.4 contains n^2 values of the form Z_iZ_j, $i = 1, \ldots, n$, $j = 1, \ldots, n$. Of these n^2 values, n will be of the form Z_i^2 and $n(n-1)$ will be of the form Z_iZ_j. However, since all the Z's consist of variables which are i.i.d their moments will be identical also. As a consequence, $E\left[Z_1^2\right] = E\left[Z_2^2\right]$ etc., and $E[Z_1Z_2] = E[Z_2Z_3]$ etc., this implies that

$$E\left[N^2\right] = nE\left[Z_1^2\right] + n(n-1)E[Z_1Z_2]. \tag{9.5}$$

Let f be any function and using the well-known fact that for any discrete random variable V we have that

$$E\left[f(V)\right] = \sum_{v\in\mathcal{V}} f(v)\,g(v)$$

where g is the probability mass function of V and $\mathcal{V}$ is the support of V. In our case $V = Z_1$ and $f(V) = Z_1^2$, $\mathcal{V} = \{0, 1\}$ and $g(1) = p$, $g(0) = 1 - p$ and this implies that

$$\begin{aligned} E\left[Z_1^2\right] &= 1^2p + \left(1-1^2p\right)\left(1-\left(1-1^2p1^2q\right)^{n-1}\right) \\ &= p + (1-p)\left(1-(1-pq)^{n-1}\right) \\ &= 1-(1-p)(1-pq)^{n-1} \\ &= E[Z_1]. \end{aligned} \tag{9.6}$$

Expanding Z_1Z_2 we get

$$Z_1Z_2 = X_1X_2 \tag{9.7}$$

$$+X_1(1-X_2)\left(1-(1-Y_{12})\Pi_{j\neq 1,2}\left(1-X_jY_{j2}\right)\right) \tag{9.8}$$
$$+X_2(1-X_1)\left(1-(1-Y_{21})\Pi_{j\neq 1,2}\left(1-X_jY_{j1}\right)\right) \tag{9.9}$$
$$+(1-X_1)(1-X_2)*$$
$$\left(\left(1-\Pi_{j\neq 1,2}\left(1-X_jY_{j1}\right)\right)\left(1-\Pi_{j\neq 1,2}\left(1-X_jY_{j2}\right)\right)\right) \tag{9.10}$$

Taking Eq. 9.7

$$E[X_1X_2]=E[X_1]E[X_2]$$
$$=p^2. \tag{9.11}$$

Taking Eq. 9.8, since only X_1 has defaulted directly this implies that X_2 must default by contagion. However this contagion cannot be caused by either X_1 or X_2 (i.e., $j \neq 1, 2$). However, since we know that X_1 has defaulted this implies the event $(1 - Y_{12})$. As a consequence we have

$$E\left[X_1(1-X_2)\left(1-(1-Y_{12})\Pi_{j\neq 1,2}\left(1-X_jY_{j2}\right)\right)\right] =$$
$$E[X_1]E[1-X_2]*E\left[1-(1-Y_{12})\Pi_{j\neq 1,2}\left(1-X_jY_{j2}\right)\right] =$$
$$p(1-p)*\left(1-(1-q)(1-pq)^{n-2}\right)$$

Taking Eq. 9.9 we obtain the same result as for Eq. 9.8.
Taking Eq. 9.10 we see that

$$\left(1-\Pi_{j\neq 1,2}\left(1-X_jY_{j1}\right)\right)*$$
$$\left(1-\Pi_{j\neq 1,2}\left(1-X_jY_{j2}\right)\right)=1-\Pi_{j\neq 1,2}\left(1-X_jY_{j1}\right)-$$
$$\Pi_{j\neq 1,2}\left(1-X_jY_{j2}\right) \tag{9.12}$$
$$+\Pi_{j\neq 1,2}\left(1-X_jY_{j1}\right)*$$
$$\Pi_{j\neq 1,2}\left(1-X_jY_{j2}\right). \tag{9.13}$$

Expanding Eq. 9.13 we get

$$\Pi_{j\neq 1,2}\left(1-X_jY_{j1}\right)*$$
$$\Pi_{j\neq 1,2}\left(1-X_jY_{j2}\right)=\Pi_{j\neq 1,2}\left(1-X_jY_{j1}\right)\left(1-X_jY_{j2}\right)$$
$$=\Pi_{j\neq 1,2}\left(1-X_jY_{j1}-X_jY_{j2}+X_j^2Y_{j1}Y_{j2}\right) \tag{9.14}$$

Taking expectations of Eq. 9.12 we get

$$E\left[1-\Pi_{j\neq 1,2}\left(1-X_jY_{j1}\right)-\Pi_{j\neq 1,2}\left(1-X_jY_{j2}\right)\right]=1-2(1-pq)^{n-2}.$$

Taking expectations of Eq. 9.14 we get

$$E\left[\Pi_{j\neq 1,2}\left(1-X_jY_{j1}-X_jY_{j2}+X_j^2Y_{j1}Y_{j2}\right)\right]=\left(1-2pq+pq^2\right)^{n-2}.$$

Taking the expectation of $(1 - X_1)(1 - X_2)$ in Eq. 9.10

$$E[(1 - X_1)(1 - X_2)] = (1 - p)^2 .$$

Combining all the above we get $E[Z_1 Z_2] = \beta_n^{pq}$. Hence substituting this value and Eq. 9.6 into Eq. 9.5 we get $E\left[N^2\right]$ and the result for the variance follows as an immediate consequence. □

As with the probability of defaults and expected number of defaults, this result allows us to define an operation *vdefaults*() : *double* of the CDO class in Fig. 9.1.

9.5 Assuming Poisson Number of Default Events

In [2] Hammarlid discusses a model in which outbreaks (i.e., the number of default events in a sector k of n_k borrowers) are considered to be a Poisson random variable Λ_k with intensity

$$\mu_k = 1 - (1 - p_k)^{n_k}$$

where p_k is the probability of default in sector k, and n_k is the number of borrowers in sector k. Figure 9.2 shows this more elaborate model of CDOs. In terms of this model, $sectors[k].mu$ represents μ_k, and $sectors[k].n$ represents n_k, etc.

The intensity corresponds to the probability of at least one borrower defaulting. Note that a single outbreak does not imply that a single borrower default has occurred, instead an outbreak denotes the occurrence of an event for which at least one borrower might default. For each outbreak in a sector k the loss due to the outbreak is $S_{kl} = N_{kl} L_k$ where N_{kl} is the number of defaults for the outbreak $l = 0, \ldots, \Lambda_k$ and L_k is the loss attributable to a default in sector k. The total loss for a sector is given by

$$S_k = \sum_{l=0}^{\Lambda_k} S_{kl}$$

and the total loss for the portfolio (assuming K sectors) is

$$S = \sum_{k=1}^{K} S_k .$$

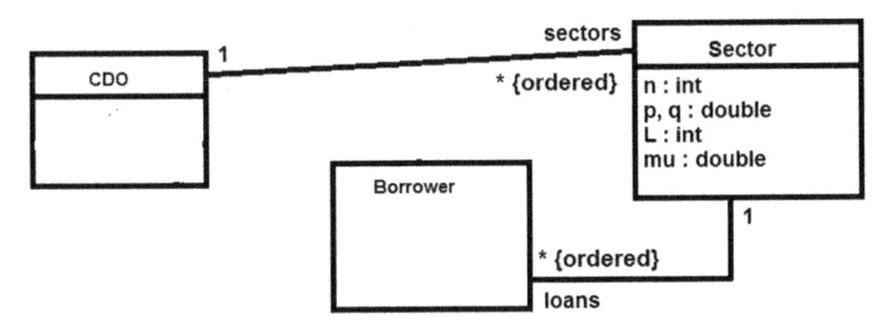

Fig. 9.2 Poisson CDO model

The key result as it relates to the above assumptions is the following theorem

Theorem 9.5.1 *Assume the number of outbreaks Λ_k to be a Poisson random variable with intensity μ_k. Then the probability $P(S = s)$ of a total loss amount s is:*

$$P(S = 0) = e^{-\sum_{k=1}^{K} \mu_k}$$
$$P(S = s) = \frac{1}{s} \sum_{k=1}^{K} \sum_{m_k=1}^{[s/L_k]} \mu_k m_k L_k P(N_k = m_k \mid N_k > 0) P(S = s - m_k L_k).$$

In the above theorem note that

$$P(N_k = m_k \mid N_k > 0) = \frac{P(N_k = m_k)}{\mu_k}. \tag{9.15}$$

Note also that $P(N_k = m_k)$ is calculated exactly as for Eq. 9.3.

Hammarlid [2] derives the moment generating function of the total credit loss:

$$M_S(\gamma) = e^{\sum_{k=1}^{K} \mu_k (\tilde{M}_k(\gamma) - 1)} \tag{9.16}$$

where

$$\tilde{M}_k(\gamma) = \frac{1}{\mu_k} \sum_{m=1}^{n_k} e^{m\gamma L_k} P(N_k = m). \tag{9.17}$$

From the above results we can define a recursive operation to evaluate $P(S = s)$ for the model of Fig. 9.2:

```
CDO::
query ps(s : int) : double
pre: s >= 0 & sectors->forAll( L > 0 )
post:
  (s = 0 =>
    result = (sectors->collect(-mu)->sum())->exp()) &
  (s > 0 =>
    result = (1.0/s)*Integer.Sum(1,sectors.size,k,vs(s,k)))
```

```
CDO::
query vs(s : int, k : int) : double
pre: s >= 0 & sectors->forAll( L > 0 )
post:
  Lk = sectors[k].L &
  result = Integer.Sum(1,(s/Lk)->floor(),mk,
                 sectors[k].vsk(mk)*ps(s-mk*Lk))
```

Here *vs* computes the inner sum in the $s > 0$ case. Because the first set of terms $\mu_k m_k L_k P(N_k = m_k \mid N_k > 0)$ are all expressions on the data of the specific *sector*$[k]$ instance, the *vs* computation can be decomposed by evaluating these expressions with an operation *vsk* of *Sector*:

```
Sector::
query vsk(mk : int) : double
post:
  result = mu*mk*L*pcond(mk)
```

where *pcond*(*mk*) computes $P(N_k = m_k \mid N_k > 0)$.

From *ps*(*s*) we can also compute *pgs*(*s*), the probability that losses are *s* or more: $P(S \geq s)$.

9.5.1 Risk Contributions in the Poisson Model

In this section we expand on the work of [2] and determine the risk contributions RC_k of each sector to the total portfolio credit loss volatility. For example, using well-known properties of moment generating functions, by twice differentiating Eq. 9.16 we can determine the expected loss and variance of the total loss distribution. First of all we deal with Eq. 9.17

$$\tilde{M}_k^{'}(\gamma) = \frac{L_k}{\mu_k} \sum_{m=1}^{n_k} mP(N_k = m)\, e^{m\gamma L_k}$$

$$\tilde{M}_k^{''}(\gamma) = \frac{L_k^2}{\mu_k} \sum_{m=1}^{n_k} m^2 e^{m\gamma L_k} P(N_k = m).$$

Now

$$\tilde{M}_k^{'}(0) = \frac{L_k}{\mu_k} \sum_{m=1}^{n_k} mP(N_k = m) \tag{9.18}$$

and

$$\tilde{M}_k^{''}(0) = \frac{L_k^2}{\mu_k} \sum_{m=1}^{n_k} m^2 P(N_k = m). \tag{9.19}$$

We also note from Eq. 9.17 that

$$\begin{aligned}\tilde{M}_k(0) &= \frac{1}{\mu_k}\sum_{m=1}^{n_k} e^0 P(N_k = m)\\ &= \frac{1}{\mu_k}\left[P(N_k = 1) + P(N_k = 2) + \cdots P(N_k = n_k)\right]\\ &= \frac{\mu_k}{\mu_k}\\ &= 1.\end{aligned}$$

The second to last equality emerges from the fact that the sum of the probabilities is equivalent to the probability of at least one borrower defaulting which is equal to μ_k.

Given the foregoing it can be observed that $M_S(0) = 1$. Further

$$M_S^{'}(\gamma) = M_S(\gamma)\sum_{k=1}^{K}\mu_k \tilde{M}_k^{'}(\gamma) \tag{9.20}$$

$$M_S^{''}(\gamma) = M_S^{'}(\gamma)\sum_{k=1}^{K}\mu_k \tilde{M}_k^{'}(\gamma) + M_S(\gamma)\sum_{k=1}^{K}\mu_k \tilde{M}_k^{''}(\gamma). \tag{9.21}$$

From the above equations, previous results and substituting 0 for γ in Eqs. 9.20 and 9.21 we see that

$$E(S) = \sum_{k=1}^{K} L_k \sum_{m=1}^{n_k} mP(N_k = m) \tag{9.22}$$

$$\begin{aligned}E\left(S^2\right) = &\sum_{k=1}^{K}\sum_{k'=1}^{K} L_k L_{k'} \sum_{m=1}^{n_k} mP(N_k = m) \sum_{m=1}^{n_{k'}} mP\left(N_{k'} = m\right) +\\ &\sum_{k=1}^{K}\sum_{m=1}^{n_k} m^2 L_k^2 P(N_k = m).\end{aligned}$$

This enables us to define an operation *expectedLoss*() : *double* of the CDO class in Fig. 9.2, which computes $E(S)$.

Given the above the variance of the total portfolio loss σ^2 is

$$\sigma^2 = \sum_{k=1}^{K}\sum_{m=1}^{n_k} m^2 L_k^2 P(N_k = m). \tag{9.23}$$

Noting that the risk contribution can be written as

$$RC_k = L_k \frac{\partial \sigma}{\partial L_k} = \frac{L_k}{2\sigma} \frac{\partial \sigma^2}{\partial L_k} \tag{9.24}$$

and using the second equality it follows that

$$RC_k = \frac{1}{\sigma} \sum_{m=1}^{n_k} L_k^2 m^2 P(N_k = m). \tag{9.25}$$

This enables us to define an operation *riskContribution*$(k : int) : double$ of CDO, quantifying the risk contribution RC_k of sector k. The higher this value, the more significant sector k is to the risk of losses for the CDO as a whole.

Given Eq. 9.25 we see that

$$\begin{aligned} \sum_{k=1}^{K} RC_k &= \frac{1}{\sigma} \sum_{k=1}^{K} \sum_{m=1}^{n_k} L_k^2 m^2 P(N_k = m) \\ &= \frac{\sigma^2}{\sigma} \\ &= \sigma. \end{aligned}$$

We can further consider individual borrowers within a sector, where each borrower has a specific loss amount L_k^i and a weighting w_k^i within the sector (*borrowers*$[i]$.*L* and *borrowers*$[i]$.*omega* in Fig. 9.3). Note that now we are considering double-valued L attributes instead of integer-valued.

If we assume that the L_k for each sector corresponds to a simple weighted average of the exposures for each individual borrower then we have a metric that would be consistent with assuming a single instantaneous default probability and contagion probability for each sector. Given this we can denote L_k in terms of its constituent actual exposures $L_k^i, i = 1, \ldots n_k$:

$$L_k = \sum_{i=1}^{n_k} w_k^i L_k^i$$

where $\sum_{i=1}^{n_k} w_k^i = 1$. Then noting that

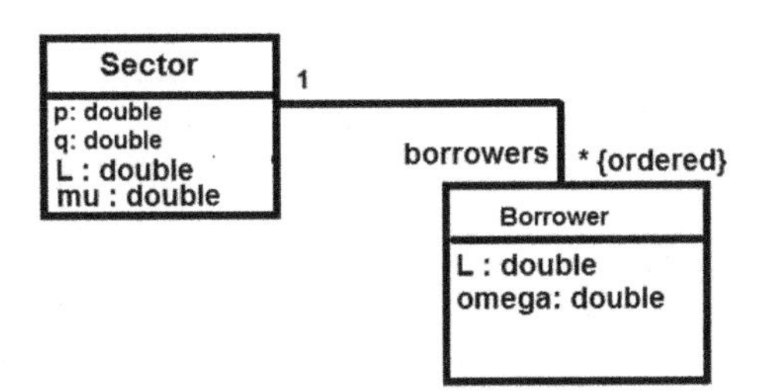

Fig. 9.3 Borrowers in sectors

$$\frac{\partial\sigma}{\partial L_k^i} = \frac{\partial\sigma}{\partial L_k}.\frac{\partial L_k}{\partial L_k^i}$$
$$= \frac{\partial\sigma}{\partial L_k}.w_k^i.$$

The last equality implies that the risk contribution of borrower i in sector k is:

$$\begin{aligned} RC_k^i &= L_k^i \frac{\partial\sigma}{\partial L_k^i} \\ &= \frac{\partial\sigma}{\partial L_k}.w_k^i L_k^i \\ &= \frac{1}{2\sigma}\frac{\partial\sigma^2}{\partial L_k} w_k^i L_k^i \\ &= \frac{1}{\sigma} w_k^i L_k^i \sum_{m=1}^{n_k} L_k m^2 P\left(N_k = m\right). \end{aligned} \tag{9.26}$$

This result enables us to define an operation

```
CDO::
query borrowerRiskContribution(k : int, i : int) : double
```

to compute RC_k^i in the Poisson CDO model.

It can also be shown that

$$\begin{aligned} \sum_{i=1}^{n_k} RC_k^i &= \sum_{i=1}^{n_k} \frac{1}{\sigma} w_k^i L_k^i \sum_{m=1}^{n_k} L_k m^2 P\left(N_k = m\right) \\ &= \frac{1}{\sigma} \sum_{i=1}^{n_k} w_k^i L_k^i \sum_{m=1}^{n_k} L_k m^2 P\left(N_k = m\right) \\ &= \frac{1}{\sigma} L_k \sum_{m=1}^{n_k} L_k m^2 P\left(N_k = m\right) \\ &= RC_{k.} \end{aligned}$$

For any probability distribution, there is a relationship between its mean, standard deviation and any percentile. Let EL, σ and Var denote the distribution expected loss, standard deviation (or volatility) of losses and the loss for a given percentile. Given a multiple ξ of σ we can define the relationship between these parameters

$$EL + \xi\sigma = Var.$$

Based on well-known properties of expectation and those we have shown already (for the decomposibility of σ into its risk contributions) we can write a variation of the above expression (as per [3]) for the contribution to the percentile or Var as

$$\hat{RC_k} = EL_k + \xi RC_k.$$

Note that, from Eq. 9.22, $\sum_{k=1}^{K} EL_k = E(S) = EL$ and $\sum_{k=1}^{K} RC_k = \sigma$. In a similar manner, and adapting Eq. 9.24, it follows that

$$\begin{aligned}\hat{RC_k^i} &= L_k^i \frac{\partial Var}{\partial L_k^i} \\ &= L_k^i \frac{\partial EL}{\partial L_k^i} + L_k^i \xi \frac{\partial \sigma}{\partial L_k^i}.\end{aligned}$$

The derivative of σ with respect to L_k^i has already been calculated and the other derivative is

$$\begin{aligned}\frac{\partial EL}{\partial L_k^i} &= \frac{\partial EL}{\partial L_k} \frac{\partial L_k}{\partial L_k^i} \\ &= w_k^i \sum_{m=1}^{n_k} mP(N_k = m).\end{aligned}$$

Putting everything together we have

$$\begin{aligned}\hat{RC_k^i} &= L_k^i w_k^i \sum_{m=1}^{n_k} mP(N_k = m) + \\ &\frac{\xi}{\sigma} w_k^i L_k^i \sum_{m=1}^{n_k} L_k m^2 P(N_k = m).\end{aligned}$$

9.6 Exact Distribution Moments

In [2] the author also derives a model for the probability distribution of losses based on the following probability generating function:

$$g(t) = \Pi_{k=1}^{K} \left(1 - \mu_k + \mu_k \hat{g}_k(t)\right) \tag{9.27}$$

where

$$\hat{g}_k(t) = \sum_{m=1}^{n_k} P(N_k = m \mid N_k > 0)\, t^{mL_k}. \tag{9.28}$$

In the following we explore the first and second order derivatives of Eq. 9.27 to determine the mean and variance of the loss distribution based on the probability generating function. The first derivative of Eq. 9.28 with respect to t, $\hat{g}_k^{'}(t)$ is:

$$\frac{\partial \hat{g}_k(t)}{\partial t} = \frac{L_k}{t} \sum_{m=1}^{n_k} mP(N_k = m \mid N_k > 0)\, t^{mL_k}. \tag{9.29}$$

The second derivative of Eq. 9.28 with respect to t, $\hat{g}_k^{''}(t)$ is:

$$\frac{\partial^2 \hat{g}_k(t)}{\partial t^2} = \frac{L_k^2}{t^2} \sum_{m=1}^{n_k} m^2 P(N_k = m \mid N_k > 0)\, t^{mL_k} - \frac{1}{t}\hat{g}_k^{'}(t). \tag{9.30}$$

It is convenient to take the log of $g(t)$:

$$\log(g(t)) = \sum_{k=1}^{K} \log\left(1 - \mu_k + \mu_k \hat{g}_k(t)\right)$$

and taking the derivative of this log expression with respect to t:

$$\frac{g^{'}(t)}{g(t)} = \sum_{k=1}^{K} \frac{\mu_k \hat{g}_k^{'}(t)}{1 - \mu_k + \mu_k \hat{g}_k(t)}$$

$$\Rightarrow$$

$$g^{'}(t) = \sum_{k=1}^{K} \frac{\mu_k \hat{g}_k^{'}(t)\, g(t)}{1 - \mu_k + \mu_k \hat{g}_k(t)}.$$

Using the quotient rule it follows that $g^{''}(t)$ is:

$$g^{''}(t) = \sum_{k=1}^{K} \frac{\mu_k \left(1 - \mu_k + \mu_k \hat{g}_k(t)\right) d\hat{g}_k^{'}(t)\, g(t) - \mu_k^2 \left(\hat{g}_k^{'}(t)\right)^2 g(t)}{\left(1 - \mu_k + \mu_k \hat{g}_k(t)\right)^2}$$

where, using the product rule:

$$d\hat{g}_k^{'}(t)\, g(t) = g(t)\left[\frac{L_k^2}{t^2} \sum_{m=1}^{n_k} m^2 P(N_k = m \mid N_k > 0)\, t^{mL_k} - \frac{1}{t}\hat{g}_k^{'}(t)\right] + \frac{L_k}{t} \sum_{m=1}^{n_k} mP(N_k = m \mid N_k > 0)\, t^{mL_k} \sum_{k=1}^{K} \frac{\mu_k \hat{g}_k^{'}(t)\, g(t)}{1 - \mu_k + \mu_k \hat{g}_k(t)}.$$

Now substituting 1 for t we see that $g(1) = 1$ and $\hat{g}_k(1) = 1$. Also

$$\hat{g}_k^{'}(1) = CEL_k = L_k \sum_{m=1}^{n_k} mP(N_k = m \mid N_k > 0)$$

where CEL_k denotes the conditional expected loss for sector k. It follows that

$$g^{'}(1) = \sum_{k=1}^{K} \mu_k \, CEL_k$$
$$= EL.$$

Note that this is the same result that was obtained for the Poisson model for the expected loss of the whole portfolio. Using the previous results the product term becomes

$$d\hat{g}_k^{'}(1)\, g(1) = \left[L_k^2 \sum_{m=1}^{n_k} m^2 P(N_k = m \mid N_k > 0) - CEL_k \right] +$$
$$L_k \sum_{m=1}^{n_k} mP(N_k = m \mid N_k > 0) \,. \sum_{k=1}^{K} \mu_k \, CEL_k$$
$$= \left[L_k^2 \sum_{m=1}^{n_k} m^2 P(N_k = m \mid N_k > 0) - CEL_k \right] + CEL_k.EL.$$

With this and the previous values substituted into $g^{''}(t)$, for $t = 1$ we get

$$g^{''}(1) = \sum_{k=1}^{K} \left\{ \mu_k \left[L_k^2 \sum_{m=1}^{n_k} m^2 P(N_k = m \mid N_k > 0) - CEL_k + CEL_k.EL \right] - \mu_k^2 \, CEL_k^2 \right\}$$
$$= \sum_{k=1}^{K} \left\{ \mu_k \left[L_k^2 \sum_{m=1}^{n_k} m^2 P(N_k = m \mid N_k > 0) - CEL_k + CEL_k.EL \right] - EL_k^2 \right\}$$
$$= \sum_{k=1}^{K} \sum_{m=1}^{n_k} m^2 L_k^2 P(N_k = m) - \sum_{k=1}^{K} EL_k + EL^2 - EL^2$$
$$= \sum_{k=1}^{K} \sum_{m=1}^{n_k} m^2 L_k^2 P(N_k = m) - EL.$$

Note that the last line of the above result is of the form:

$$g_X^{''}(1) = E\left(X^2\right) - E(X).$$

However, using a basic property of probability generating functions we know that

$$E\left(X^2\right) - E(X) = Var(X) + [E(X)]^2 - E(X)$$
$$\Longrightarrow Var(X) = g_X^{''}(1) - [E(X)]^2 + E(X).$$

Applying the above to the $g^{''}(1)$ we get

$$\sigma^2 = \sum_{k=1}^{K} \sum_{m=1}^{n_k} m^2 L_k^2 P\left(N_k = m\right) - \mu_k^2\, CEL_k^2.$$

As in the case of the risk contributions for the Poisson model we know that

$$\begin{aligned} RC_k &= \frac{L_k}{2\sigma} \frac{\partial \sigma^2}{\partial L_k} \\ &= \frac{1}{\sigma}\left[\sum_{m=1}^{n_k} L_k^2 m^2 P\left(N_k = m\right) - \mu_k^2\, CEL_k^2\right]. \end{aligned}$$

Summing these risk contributions over all sectors we get

$$\begin{aligned} \sum_{k=1}^{K} RC_k &= \frac{1}{\sigma} \sum_{k=1}^{K} \sum_{m=1}^{n_k} L_k^2 m^2 P\left(N_k = m\right) - \mu_k^2\, CEL_k^2 \\ &= \frac{\sigma^2}{\sigma} \\ &= \sigma. \end{aligned}$$

Based on similar arguments as in the Poisson case it can be shown that

$$RC_k^i = w_k^i L_k^i \frac{1}{\sigma}\left[\sum_{m=1}^{n_k} L_k m^2 P\left(N_k = m\right) - \mu_k^2\, CEL_k^2\right]$$

and clearly that $\sum_{i=1}^{n_k} RC_k^i = RC_k$.

9.7 Allocating Borrowers Across Sectors

The analysis conducted so far assumes that contagion only occurs within sectors and that borrowers only belong to a single sector. We now relax the assumption that a borrower can only belong to a single sector. By assuming that borrowers can be associated with more than one sector we allow for the simultaneous contagion of several sectors by the same borrower and hence we introduce a form of inter-sector contagion, including for example the case of a business with activities in several sectors. In the sequel we focus attention on the Poisson model.

Let θ_k^a denote the percentage allocation of borrower a to sector k. This means that one borrower, b, may have an allocation of (for example), 0.5 to one sector, 0.3 to another and 0.2 to a third. This can be represented by further extending the CDO model to include an intermediary entity between *Sector* and *Borrower* (Fig. 9.4).

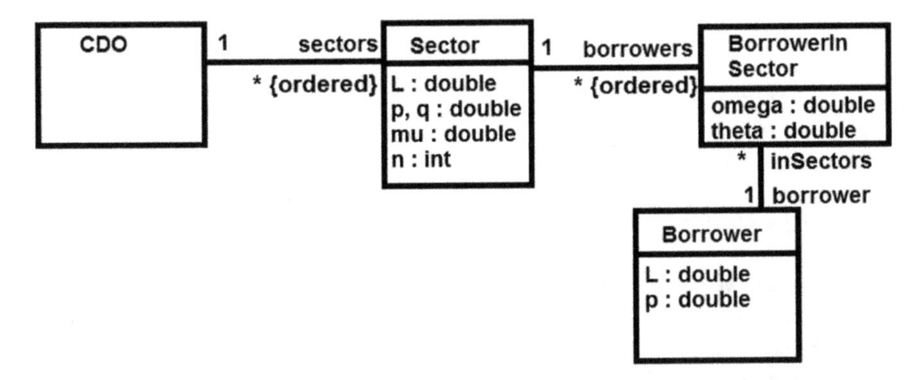

Fig. 9.4 Borrowers in multiple sectors

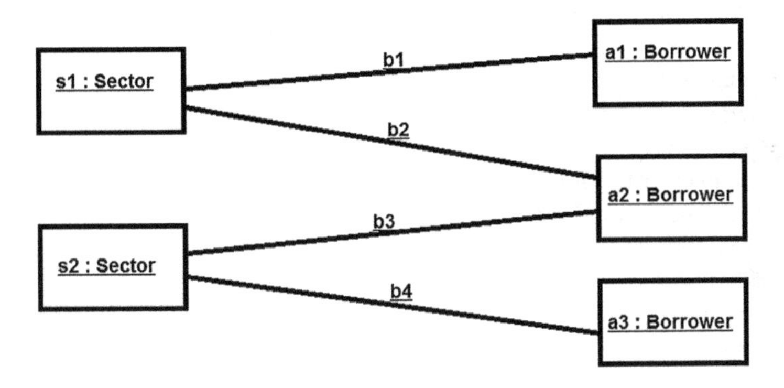

Fig. 9.5 Example of borrowers in multiple sectors

To visualise this situation, consider an example where there are three borrowers $a1$, $a2$, $a3$, and two sectors, $s1$, $s2$ (Fig. 9.5). The *BorrowerInSector* objects correspond to the links between the sectors and borrowers.

As a text model, the elements could be written as:

```
c : CDO
s1 : Sector
s1 : c.sectors
s1.n = 2
s2 : Sector
s2 : c.sectors
s2.n = 2

a1 : Borrower
a1.L = 5
a1.p = 0.02
a2 : Borrower
```

```
a2.L = 7
a2.p = 0.01
a3 : Borrower
a3.L = 4
a3.p = 0.03
```

The linking elements could be:

```
b1 : BorrowerInSector
b1.borrower = a1
b1.theta = 1
b1.omega = 0.6
b1 : s1.borrowers
b2 : BorrowerInSector
b2.borrower = a2
b2.theta = 0.7
b2.omega = 0.4
b2 : s1.borrowers
b3 : BorrowerInSector
b3.borrower = a2
b3.theta = 0.3
b3.omega = 0.5
b3 : s2.borrowers
b4 : BorrowerInSector
b4.borrower = a3
b4.theta = 1
b4.omega = 0.5
b4 : s2.borrowers
```

This means that 60% of $s1$ consists of $a1$ and 40% of $a2$, whilst $s2$ has equal proportions of $a2$ and $a3$. $a2$ is split between the sectors, 70% in $s1$ and 30% in $s2$.

The θ_k^a for a given borrower a should have the properties that:

$$\sum_{k=1}^{K} \theta_k^a L_a = L_a$$

$$\sum_{k=1}^{K} \theta_k^a p^a = p^a$$

$$\sum_{k=1}^{K} \theta_k^a = 1$$

where L_a denotes the total exposure to borrower a in a portfolio of borrowers (i.e., across all sectors) and p^a denotes the instantaneous default probability for the borrower.

In terms of Fig. 9.4 this means that:

$$inSectors \rightarrow collect(theta) \rightarrow sum() = 1$$

for each borrower.

The *omega* weightings represent the relative contributions of borrowers within a given sector, thus:

$$borrowers \rightarrow collect(omega) \rightarrow sum() = 1$$

for each sector.

From Eq. 9.23 we recall that

$$\sigma^2 = \sum_{k=1}^{K} \sum_{m=1}^{n_k} m^2 L_k^2 P\left(N_k = m\right).$$

However due to the introduction of the partial allocation to a sector we have that L_k is a weighted sum of the (partial) losses due to the borrowers in sector k:

$$L_k = \sum_{i=1}^{n_k} w_k^i \theta_k^i L_i$$

In terms of the model of Fig. 9.4, this means

$$\begin{aligned}&Sector ::\\ &\quad L = borrowers \rightarrow collect(omega * theta * borrower.L) \rightarrow sum()\end{aligned}$$

Similarly for p. In our example, $s1.L$ can be calculated to be 4.96.

Therefore:

$$\sigma^2 = \sum_{k=1}^{K} \sum_{m=1}^{n_k} m^2 \left(\sum_i (w_k^i \theta_k^i)^2 L_i^2 + 2 \sum_{i<j} w_k^i w_k^j \theta_k^i \theta_k^j L_i L_j \right) P\left(N_k = m\right).$$

The risk contributions can now be calculated

$$\begin{aligned}
L_a \frac{\partial \sigma}{\partial L_a} &= \frac{L_a}{2 * \sigma} \frac{\partial \sigma^2}{\partial L_a} = RC_a \\
&= \tfrac{L_a}{2\sigma} \sum_{k=1}^{K} \sum_{m=1}^{n_k} m^2 \left(2(w_k^a \theta_k^a)^2 L_a + 2 \sum_j w_k^a w_k^j \theta_k^a \theta_k^j L_j \right) P\left(N_k = m\right) \\
&= \frac{1}{\sigma} \sum_{k=1}^{K} \sum_{m=1}^{n_k} m^2 \left((w_k^a \theta_k^a)^2 L_a^2 + \sum_j w_k^a w_k^j \theta_k^a \theta_k^j L_j L_a \right) P\left(N_k = m\right).
\end{aligned}$$

Summing over all borrowers a we have

$$\begin{aligned}
\sum_a RC_a &= \frac{1}{\sigma} \sum_a \sum_{k=1}^{K} \sum_{m=1}^{n_k} m^2 \left((w_k^a \theta_k^a)^2 L_a^2 + \sum_j w_k^a w_k^j \theta_k^a \theta_k^j L_j L_a \right) P(N_k = m) \\
&= \frac{1}{\sigma} \sum_{k=1}^{K} \sum_{m=1}^{n_k} m^2 \left(\sum_a (w_k^a \theta_k^a)^2 L_a^2 + \sum_a \sum_j w_k^a w_k^j \theta_k^a \theta_k^j L_j L_a \right) P(N_k = m) \\
&= \frac{\sigma^2}{\sigma} \\
&= \sigma.
\end{aligned}$$

Based on earlier results it can be shown that

$$\begin{aligned}
L_a \frac{\partial Var}{\partial L_a} &= L_a \frac{\partial EL}{\partial L_a} + L_a \xi \frac{\partial \sigma}{\partial L_a} \\
&= L_a \sum_{k=1}^{K} \sum_{m=1}^{n_k} m P(N_k = m)\, w_k^a \theta_k^a \\
&\quad + \frac{\xi}{\sigma} \sum_{k=1}^{K} \sum_{m=1}^{n_k} m^2 \left((w_k^a \theta_k^a)^2 L_a^2 + \sum_j w_k^a w_k^j \theta_k^a \theta_k^j L_j L_a \right) P(N_k = m).
\end{aligned}$$

Summary

In this chapter we have illustrated the process of mathematical analysis of financial products by considering the properties of CDOs in depth. Alongside the mathematical derivations we have provided the UML models and the OCL formalisations of the mathematical equations.

In general, it is more effective to use conventional mathematical notation to derive theoretical results using the standard tools of statistics, probability theory and calculus, and then use OCL and UML as a bridge between the mathematics and practical computation.

Exercises

1. Formalise the *vdefaults*() operation of the simple CDO version (Fig. 9.1).
2. Formalise the *expectedLoss*() operation of the Poisson CDO version (Fig. 9.2).

3. In the Poisson CDO version, assume there is one sector with N borrowers, with loss amount $L = 3$. What can be deduced about $P(S = 1)$ or $P(S = M)$ where M is not a multiple of 3?

4. Define $pgs(s)$ in terms of $ps(s)$ for the Poisson CDO model.

5. Formalise the $riskContribution(k : int)$ operation of the Poisson CDO version.

6. Generalising loss amounts L to be double-valued also implies the need to generalise $P(S = s)$ to double-valued s. Define additional clauses of ps and vs to handle this extension.

7. Formalise the $borrowerRiskContribution(k : int, i : int)$ operation of the Poisson CDO version.

8. Compute L and p for the two sectors in the example model of Fig. 9.4.

References

1. M. Davis, V. Lo, Infectious defaults. Quant. Financ. **1**(4), 382–387 (2001)
2. O. Hammarlid, Aggregating sectors in the infectious defaults model. Quant. Financ. **4**(1), 64–69 (2004)
3. Credit Suisse First Boston, CreditRisk+, A Credit Risk Management Framework (1997)

Chapter 10
Tool Support for Financial Application Development

In this chapter we describe in detail how to use the UML-RSDS tools to specify and implement financial applications on a number of different platforms. We describe how the tools can be combined with the use of Excel, and how the tools can be extended to provide code-generation facilities for new target programming languages.

10.1 Using UML-RSDS

UML-RSDS provides tools to create and edit UML class diagrams, use case diagrams and other UML notations. The class diagram describes the data of an application, whilst the use cases define the services which the application offers. We combine these two views into a single integrated model, in order to enhance agility.

The tools can be obtained from www.nms.kcl.ac.uk/kevin.lano/uml2web or from Eclipse, they run under Java on either Windows, Mac or Unix platforms.

10.2 Case Study: Extended Bootstrapping

To illustrate the process of specification using UML-RSDS, we revisit the interest rate bootstrapping example of Sect. 5.3. Currently, the application takes in a series of *known* interest rates as a parameter. But it is more usual for finance practitioners to use Excel spreadsheets or plain text data tables to store source data for computations. The bootstrap application can be used in this way by introducing a new class, *YieldData*, which represents a data table of known interest rates (Fig. 10.1).

Objects of the *YieldData* class correspond to rows of the table. Tables normally have a primary key column, this corresponds to an identity attribute in UML-RSDS, and these must be of String type. In this application, *maturity* is the primary key for

K. Lano and H. Haughton, *Financial Software Engineering*, Undergraduate Topics in Computer Science, https://doi.org/10.1007/978-3-030-14050-2_10

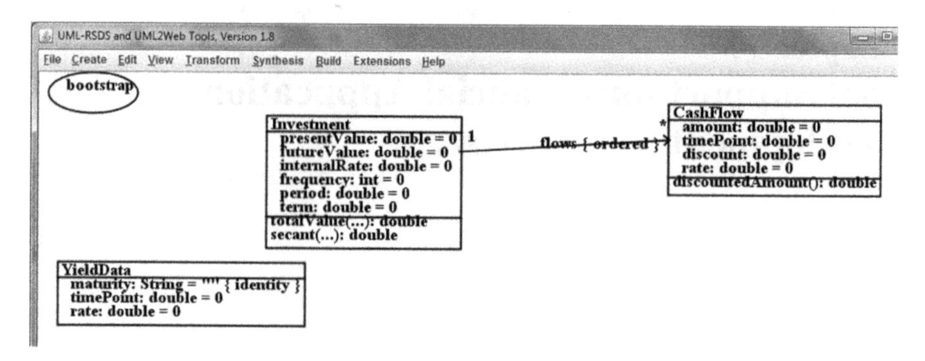

Fig. 10.1 Bootstrap with data table

YieldData: this is the String equivalent of *timePoint*. *rate* then gives the yield for that maturity. Data of instances can be read from and written to a file *YieldData.csv* in text spreadsheet format. This means that the generated application can be used in an integrated manner with existing spreadsheet applications.

An example table could be:

```
"1", 1, 0.017
"2", 2, 0.015
"3", 3, 0.018
```

This defines three known interest rates (yields) for investments of duration 1, 2, and 3 years respectively. The order of the columns corresponds to the order of the attribute declarations within the class.

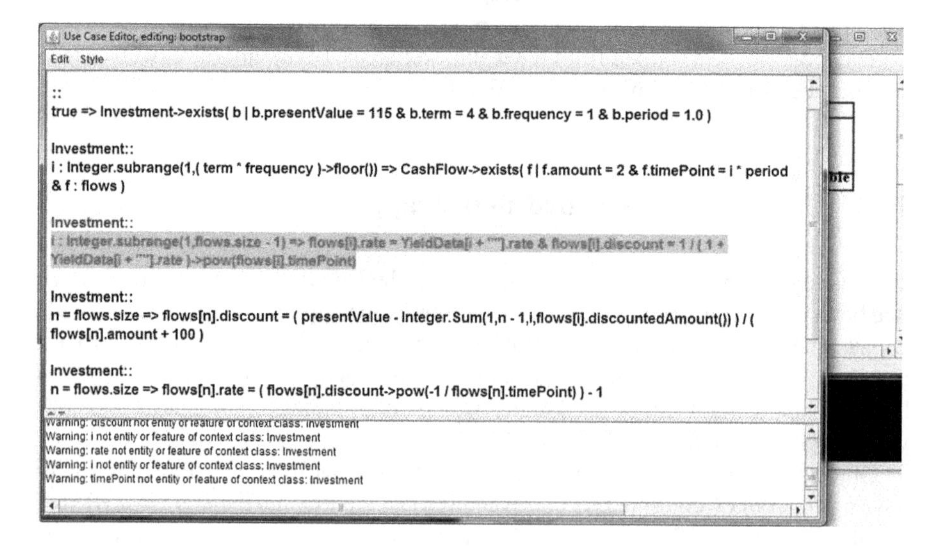

Fig. 10.2 Use case editor

Within the application, the rate for integer duration t is obtained as $YieldData[t + ""].rate$. This can be used instead of $known[t]$ in the specification.

Editing of the use case to make this change can be carried out using the "Edit use case" option on the Edit menu, which opens up a specialised editor (Fig. 10.2).

The constraints can each be checked before saving the use case. Errors are highlighted in red and warnings and error messages are listed in the lower window. Once the specification is syntactically correct, it can be type-checked, and then a design synthesised from it, using the options on the Synthesis menu.

Finally, using the Build menu options, executable code in different languages can be generated, to enable testing of the application. Detected errors should be addressed by editing the *specification*, not the generated code.

10.3 Code Generation

Automated code generation from high-level models is a key to performance gains from model-driven engineering. Instead of time-consuming and expensive manual programming, executable code can be automatically synthesised from relatively concise and simple descriptions of system behaviour. For financial applications, the specification in OCL notation is usually similar to the mathematical equations that are used to define financial models. Hence it is relatively direct to express the financial mathematics in UML-RSDS and to validate the correctness of the specification by inspection.

Complex code mechanisms such as design patterns, caching and maintaining mutual consistency of different data items can be automatically constructed by a code generator, thus reducing the possibility of code errors. The code produced will usually have a consistent structure, facilitating understanding, testing and integration.

However, code generators may produce excessively long and complex code containing redundancy and duplication, which a skilled human programmer would avoid. Bugs in the generator may still exist in rarely-used situations, even if the generator has been used repeatedly without problems for more common cases.

What should an application developer do if they uncover such a flaw in a code generator? In most cases they are forced into a workaround using manual coding to patch the flaw, because the code generator itself cannot be changed (it may not be open source, or the application developer may not have the necessary knowledge to modify it). We consider that a specialised MDE tooling team should be responsible for such tool support work. In addition, we publish the specification of our code generators, so that they can, if necessary, be adapted and modified *using the same process as for the development of general applications*.

For example, the UML-RSDS to Python code generator is defined in the file uml2pymm.txt in www.nms.kcl.ac.uk/kevin.lano/libraries. This generates Python code in a file app.py which uses the Python library ocl.py for OCL data types and operations. As yet, there is no facility to read in data from CSV files, as described above (and provided for Java, C# and C++). Such a facility could be added to the Python code generator by defining code to generate an operation

```
def loadCSVFiles():
## for each defined entity, E,
## look for file E.csv and create
## E instances for each line in the file
```

This is a self-contained and moderately complex task which should not take more than one day to complete.

Such code-generation tasks usually involve:

- Investigating what the code in the target language should be—having regard to simplicity and efficiency
- Identifying what information in the source metamodel (such as UML) is needed to produce the required code
- Organising the code generation task within the overall code generation process.

In the present case, investigation of Python file-processing facilities identifies that the code for reading a table *E.csv* of data will have the form:

```
Efile = open('E.csv', 'r')
for line in Efile :
  ex = E()
  values = tokeniseCSVLine(line)
  ex.att1 = conversionatt1(values [0])
  ...
  ex.attn = conversionattn(values[n-1])
Efile.close()
```

Where *tokeniseCSVLine* is a library function that splits the line on ',' symbols, but also takes account of strings, so that ',' within a string is ignored. This function can be defined as a specific routine in ocl.py. The conversion function *conversionatti* of a string value into the appropriate attribute type for *atti* depends on information in the source metamodel. We need information from the *Entity*, *Property* and *Type* metaclasses of the UML-RSDS metamodel (Fig. 10.3):

- We need a loadCSV operation for each *leaf* class *E* : *Entity* (only leaf classes can be concrete in UML-RSDS)
- A row cell must be processed for each *ownedAttribute* of such *E* which is of *PrimitiveType* (including numerics, booleans and strings), and for all such inherited attributes
- The conversion function required is *int*(*value*) for *int* and *long*-valued attributes, *float*(*value*) for *double*-valued, and an empty conversion for strings.

In terms of organising the code production, the *loadCSV* operations can be generated after the creation operations and before the operations for use cases, in app.py.

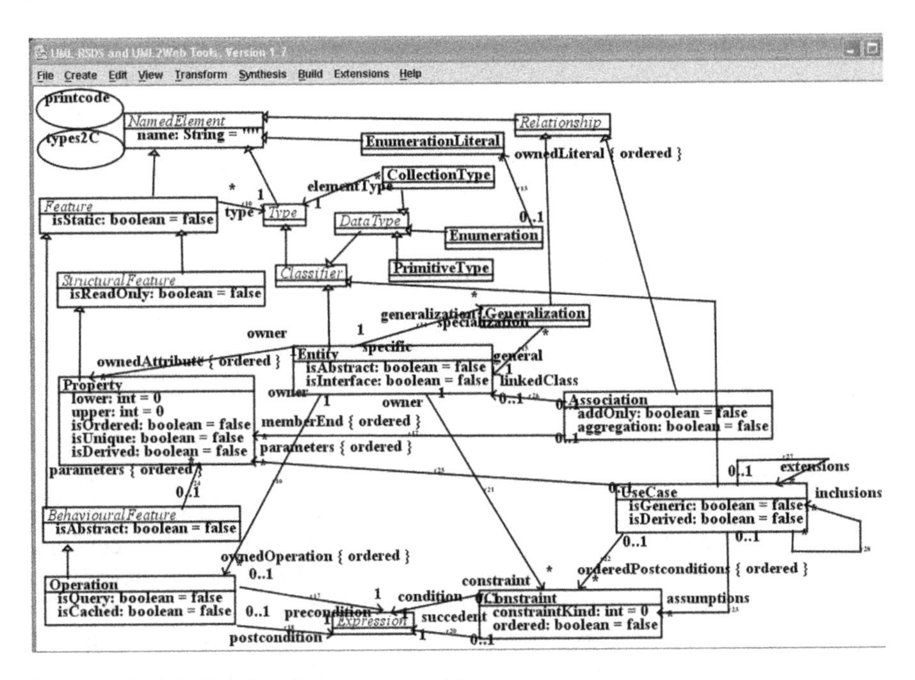

Fig. 10.3 UML-RSDS class diagram metamodel

The new specification text in uml2pymm.txt is therefore of the form:

```
Entity::
  isLeaf &
  atts = allAttributes()->select( type : PrimitiveType)  &
  ex = name.toLowerCase + "x" =>
   (name + "file = open('" + name + ".csv', 'r')")->display() &
   ("for line in " + name + "file :")->display() &
   ("   " + ex + " = " + name + "()")->display() &
    "   values = tokeniseCSVLine(line)"->display() &
    Integer.subrange(1,atts.size)->forAll( i |
      ("   " + ex + "." + atts[i].name + " = " +
          atts[i].conversion() +
          "(values[" + (i-1) + "])")->display()) &
    (name + "file.close()")->display()
```

The specification is based closely on the required code (i.e., it is a *template* of the code), with some variable elements based on source metamodel information. The *conversion* operation is added to uml2pymm.txt as an operation of *Property*.

10.4 Creating New Code Generators

The UML-RSDS tools can be used to create new DSLs and associated tools that generate artifacts, including code, from DSL models. They can also be used to write code generators for new target languages.

To define a new code generator from UML-RSDS to a programming language, the development can be organised into five main parts:

- F1.1: Translation of types
- F1.2: Translation of class diagrams
- F1.3: Translation of OCL expressions
- F1.4: Translation of activities
- F1.5: Translation of use cases.

For each part an informal mapping of language elements from UML to the target language is first constructed. For each UML language element (i.e., each metaclass in Fig. 10.3), the concrete syntax of its representation in the target language is identified.

The type mapping is fundamental, since types are used in all other parts of UML-RSDS: to give types to class features and to expressions, and to expressions within statements and use cases. Thus this mapping should be established first since it determines aspects of the other mappings. Table 10.1 shows the informal mapping of types for UML to Python.

For the mapping of class diagrams, F1.2, we identify how classes and their owned properties and operations should be represented in Python, including a representation of inheritance. Table 10.2 gives an informal mapping for these elements.

This mapping reflects a decision to utilise Python's dynamic typing to express UML inheritance.

A similar specification can be defined for mapping expressions (F1.3), activities (F1.4), based on the activity metamodel (Fig. 3.7), and use cases (F1.5). The detailed definitions of these clauses will depend on the target programming language, however

Table 10.1 Informal mapping scenarios for UML types to Python

Scenario	UML element e	Python representation e'
F1.1.1.1	*String* type	`str`
F1.1.1.2	int, long, double types	`int, int, float`
F1.1.1.3	boolean type	`bool`
F1.1.2	Enumeration type	Enum class instance
F1.1.3	Entity type *E*	class `E`
F1.1.4.1	*Set*(*E*) type	`set`
F1.1.4.2	*Sequence*(*E*) type	`list`
F1.1.4.3	*Map*(*K*, *T*) type	`dict`

Table 10.2 Informal scenarios for the mapping of UML class diagrams to Python

Scenario	UML element e	C representation e'
F1.2.1	Class diagram *D*	Python program with *D*'s name

F1.2.2	Class/interface *E*	`class E` : definition
		static variable `e_instances = []`
		`def __init__(self) :`
		
		`E.e_instances.append(self)`
		`def createE()` : operation

F1.2.3.1	Instance-scope attribute $p : T$	Instance variable `p = T'Init` defined in *init* of *E*
F1.2.3.2	Principal identity attribute p : *String* of class *E*	Static `e_index = dict({})` variable of *E* `def getEByPK(v)` : operation `def createByPKE(v)` : operation
F1.2.3.3	class-scope attribute *att* : *T*	static variable `att = T'Init`
F1.2.4	Operation *op*($p : P$) : *T* of *E* (instance-scope)	Python operation `def op(self,p) :`
F1.2.5	Inheritance of *A* by *B*	Features of *A* are listed in `class B`

the structure of the specification and the cases to be considered will always be the same. Thus it is possible to copy a generator such as *uml2py* and adapt it to a new target language, e.g., JavaScript or Matlab.

10.5 Defining Domain-Specific Languages (DSLs)

New software languages can also be defined using UML-RSDS, together with supporting tools. For example, consider the JSON (JavaScript Object Notation) text format for structured data [1]. This is typically used to transfer small amounts of data between server and client in a web interaction, or as a source format for structured data of any kind. An example of JSON data is:

```
[{ "share" : "IBM",
   "date" : "2018:12:28", "opening" : "123.6",
   "closing" : "122.1", "volume" : "110000"},
 { "share" : "MSFT",
   "date" : "2018:12:28", "opening" : "153.0",
   "closing" : "155.2", "volume" : "199000"}]
```

describing some share data. The JSON data has the form of a list of objects, where each object is a list of pairs “key” : “value” giving the value of each property of the object.

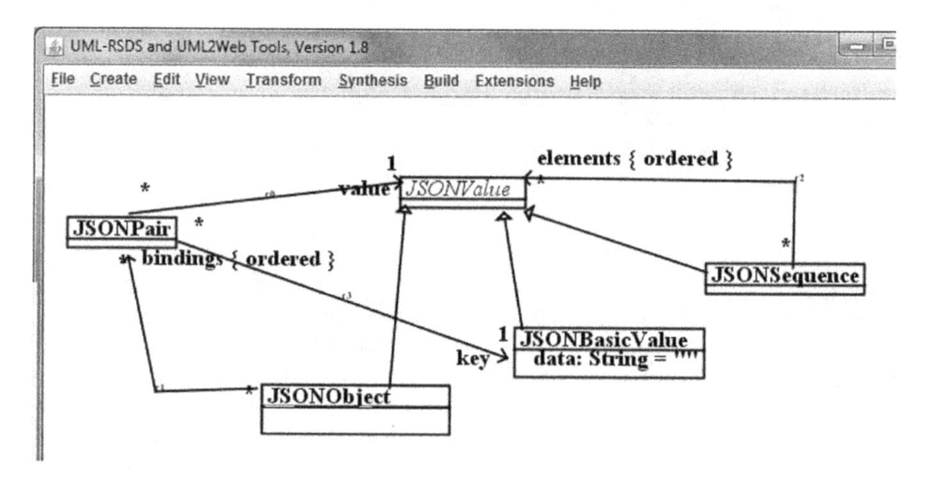

Fig. 10.4 JSON metamodel

We may need to transform such data into many different forms, such as XML (for FIX processing), HTML (to display on a web page), plain text, program language data declarations, or code to store the data in a database, etc.

The steps needed to define a DSL and tools in UML-RSDS are:

1. Define a class diagram to formalise the concepts of the language. This defines the abstract syntax of the DSL.
2. Define a concrete syntax for the DSL, or adopt one if such a syntax already exists.
3. Write UML-RSDS operations and use cases to formalise different tools to operate on DSL models.
4. Use the UML-RSDS tool to build stand-alone tools for the DSL from the tool specifications. These will be Java .jar files.

For JSON, these steps become:

1. Define the class diagram of the language metamodel (Fig. 10.4).
2. Adopt the existing concrete syntax for JSON.
3. Specify required tools, such as a JSON to HTML transformation, in UML-RSDS.
4. Code-generate the tools using the *Synthesis* and *Build* options.

Because of the recursive nature of the JSON format, the natural form of specifications processing JSON are operations for each metaclass, that typically invoke other operations to process their sub-parts. For example, printing the concrete syntax of JSON data can be achieved by the operations:

```
JSONValue::
query abstract toString() : String

JSONPair::
```

```
query toString() : String
post:
  result = key + " : " + value

JSONObject::
query toString() : String
post:
  (bindings.size = 0 => result = "{}") &
  (bindings.size = 1 =>
     result = "{ " + bindings [1] + " }") &
  (bindings.size > 1 =>
     result = "{ " + bindings [1] +
       bindings.tail->collect( p | ", " + p )->sum() + " }")

JSONBasicValue::
query toString() : String
post:
  result = data

JSONSequence::
query toString() : String
post:
  (elements.size = 0 => result = "[]") &
  (elements.size = 1 => result = "[ " + elements[1] + " ]") &
  (elements.size > 1 =>
     result = "[ " + elements[1] +
       elements.tail->collect( p | ", " + p )->sum() + " ]")
```

The application can be edited either using the graphical class diagram editor, or equivalently in the text KM3 editor (Fig. 10.5).

In a similar manner, a mapping to HTML tables could be defined as:

```
JSONValue::
query abstract toHTML() : String

JSONPair::
query toHTML() : String
post:
  result = key.data + " : " + value.toHTML()

JSONObject::
query toHTML() : String
post:
  (bindings.size = 0 => result = "") &
  (bindings.size >= 1 =>
```

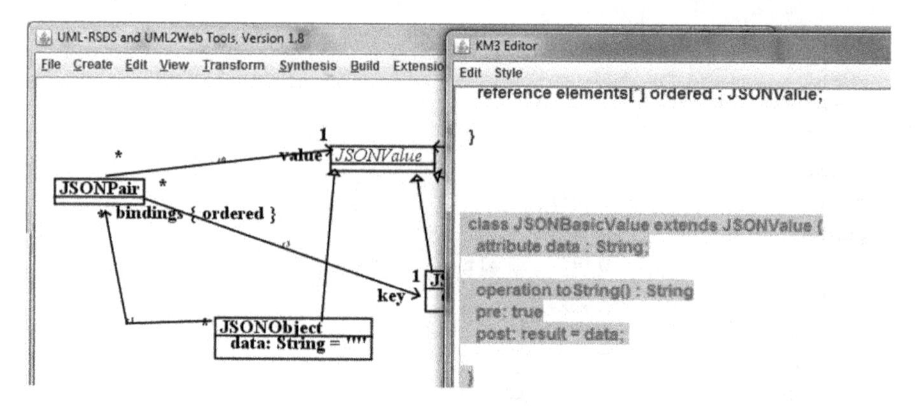

Fig. 10.5 KM3 editor

```
      result = bindings->collect( p |
        "<td> " + p.toHTML() + " </td>" )->sum())

JSONBasicValue::
query toHTML() : String
post:
  result = data

JSONSequence::
query toHTML() : String
post:
   (elements.size = 0 => result = "") &
   (elements.size >= 1 =>
      result = "<table> " + elements->collect( p |
         "<tr> " + p.toHTML() + " </tr>\n")->sum() +
               " </table>")
```

Notice that in this specification we need to call *toHTML* explicitly, whilst in the *toString* specification there are implicit recursive calls. The assumed structure of the JSON document is a sequence of objects, which becomes an HTML table with a row for each object.

10.6 Libraries

A set of libraries are defined that provide mathematical and financial functions, together with necessary data structures such as vectors and matrices. These can be imported into UML-RSDS specifications.

The libraries are:

- *FinLib*, with functions *euPutOptionPrice*, *euCallOptionPrice*
- *Real*, with operations
  - *subrange(low : double, step : double, upper : double) : Sequence(double)* to provide sequences of doubles ranging from *low* to *upper* in intervals *step*
  - *minValue() : double*, *maxValue() : double* which return the minimum and maximum permitted double values.
- *MathLib*, with operations *random*, *factorial*, *combinatorial*, *gcd*, *lcm*, *pi*, *e*, *isPrime*, and various hyperbolic, conversion and integration functions.
- *NormalDist*, with operations *normal*, *cumulative*, *sample* for the normal distribution.
- *Matrix*, with a wide range of operations on (2-dimensional) matricies.
- *Sequences*, with operations on vectors, considered as sequences.
- *StatLib*, with operations on statistical distributions, such as
  - *mean(Sequence(double)) : double*
  - *standardDeviation(Sequence(double)) : double*, etc.
- *NumericOptLib*, with operations for *secant*, *bisection* and other numerical optimisations. This is linked to classes *Simplex* and *SimplexPoint* for the simplex algorithm.
- *StringLib*, with operations *before(str : String, delim : String) : String*, *after(str : String, delim : String) : String*, *split(str : String, delim : String) : Sequence (String)*, etc.
- XMLParser, to parse XML documents.

These libraries are defined in *realmm.txt*, *mathlibmm.txt*, etc, at https://nms.kcl.ac.uk/kevin.lano/uml2web/libraries

Summary

In this chapter we have given details of the UML-RSDS process and tools to support synthesis of financial application from specifications. We have given guidance on adapting and extending the tools to provide additional functionality.

Reference

1. JavaScript Object Notation, https://www.w3schools.com/js/js_json_intro.asp (2018)

Appendix A
Glossary

Basis Point

0.01 of a percentage point. Often used in relating different rates of interest, to express that one rate is X basis points above or below another.

Compounding

The combination of a capital sum with an increment (or decrement) due to an interest rate. *Discrete compounding* involves the increment being applied annually or a fixed finite number of times per year. The equation is

$$Y = X * (1 + r)^N$$

for annual compounding of X at rate r for N years, and

$$Y = X * \left(1 + \frac{r}{f}\right)^{f*N}$$

for frequency f-per-year compounding of X at annual rate r for N years. *Continuous compounding* increases the frequency f to infinity, and mathematically this results in the equation

$$Y = X * e^{r*N}$$

Credit Crunch

The financial crisis of 2008–2009, caused by defaults in securities backed by sub-prime mortgages, led to widespread losses throughout the banking system, and hence to an economic slowdown which still affects the UK economy.

Derivative Security

A security whose value is based upon the value of other securities or assets.

K. Lano and H. Haughton, *Financial Software Engineering*, Undergraduate Topics in Computer Science, https://doi.org/10.1007/978-3-030-14050-2

Hedging

Reducing the risk of losses in a main position by creating an alternative position which will compensate for adverse results in the main position.

For example, using options to take an opposite position in the market to a main position.

In the Money

A contract is *in the money* to an investor if it represents a profit for them given the current values of assets. For example, an American option on a stock can be exercised at any time (in principle) and such a call option would be in the money at any time where the stock price is greater than the option strike price.

A contract is *at the money* if it is in a state of zero profit/loss currently, and *out of the money* if it represents a current loss.

Leverage Ratio

The ratio of a businesses liabilities to its equity. E.g., a business with assets of £1.05 million and liabilities of £1 million has equity of £50,000 and a leverage ratio of 20.

Leveraging

Using a relatively small amount of funds to obtain the effect of a larger investment. For example, speculating on the price of a share rising above price P by buying call options in the share with strike price P, instead of buying actual shares. The option price may be only 5% of the share price: a leverage ratio of 20.

Liquidity

The degree to which funds can be transferred from one asset to another or realised in cash. Real-estate is an example of an illiquid asset, whilst shares are usually a highly liquid asset.

Long Position

The party in a contract who agrees to buy an asset.

Net Present Value (NPV)

The income from an investment, minus the amounts invested, all costed in terms of current values. That is, a cash flow of amount X to be received in N years time is discounted to $X/(1+r)^N$ or $X * e^{-r*N}$ where r is the annual interest rate that applies over this time period.

Over-the-Counter Market

Trading of securities directly between organisations without the mediation of an exchange.

Repo (Repurchase) Rate

A risk-free interest rate r based on repurchase agreements for securities: the owner of securities obtains a loan based on temporary transfer of ownership of the securities to the lender.

Short Position

The party in a contract who agrees to sell an asset.

Value at Risk (VaR)

A risk measure which quantifies the maximum amount of loss L that can arise in an investment over a specified time period (e.g., one day) within a probability bound. I.e.:

$$L > VaR_p \implies prob(Loss \geq L) \leq p$$

For example, a one-day 5% VaR of £500,000 means that over one day, there is a 5% chance that losses will be greater than this amount, and a 95% chance that losses will be less. VaR is commonly used for risk management and measurement in the finance industry.

Although VaR is intuitively clear to understand, there are some theoretical objections to this measure, and concerns that it does not address the risks of low-frequency but high-impact events.

Yield

The yield of an investment measures the rate at which it returns value to the investor. It is the unique or smallest interest rate r at which the net present value of the investment is 0, in other words, the rate at which the investment breaks even.

Yield can be used as a guide to the quality of an investment and hence guide investment decisions. Other factors, such as risk levels (high risk and high yields tend to go together) and the term of an investment are also important.

Yield Curve

A graph of the yield (y-axis) against the duration (x-axis) of zero-coupon investments. This is usually upwards-sloping, indicating that investors prefer the greater liquidity of short-term investments, and hence higher yields need to be offered for longer-term investments. In addition investors require a greater rate of return for the increased uncertainty of lending over longer periods.

Appendix B
Exercise Solutions

Chapter 2

1. This is the *payout* of the bond.

2. The bond with price £105 will have the higher yield, because if $105 = value(r)$ and $110 = value(r')$ on the same bond data, then $r' < r$.

3. It converges to the redemption amount, 100.

4. Compounding an amount X for t years now involves multiplying X by $1 + rs[i]$ for $i = 1$ to t. Therefore:

$$value(rs) = \left(\Sigma_{t=1}^{term} coupon/\Pi_{i=1}^{t}(1 + rs[i])\right) + 100/\Pi_{i=1}^{term}(1 + rs[i])$$

5. The β_1 parameter controls the height of the curve and a change of δ in this parameter shifts the curve up or down by this amount.

6. Several situations could lead to default infection between sectors: (i) a banking failure in the financial sector would adversely affect any business that holds funds in the bank; (ii) a failure of a major company can cause the failure of companies which depend on the company for business, e.g., companies in the supply chain of the failed company, such as aero engine suppliers which depend on a specific airline for sales; (iii) a company with subsidiaries in different sectors, which could go out of business if the parent company does.

7. Here the accrued interest is 1.45329, DSC/E is 0.4945, A/E is 0.5055, resulting in a value of 96.088.

Chapter 4

1. There would be a $0 \ldots 3$ multiplicity on the *customers* role end, instead of $*$.

K. Lano and H. Haughton, *Financial Software Engineering*, Undergraduate Topics in Computer Science, https://doi.org/10.1007/978-3-030-14050-2

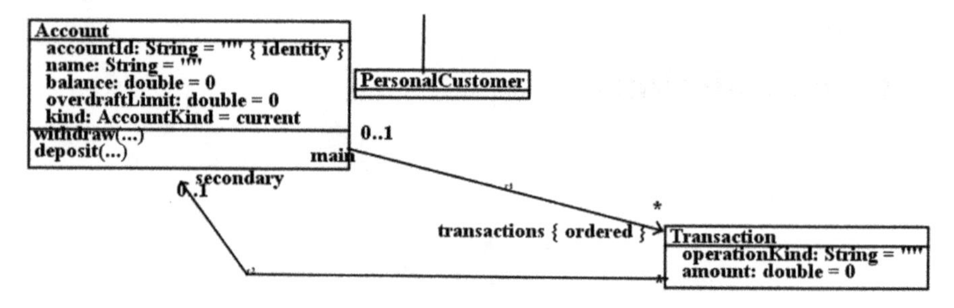

Fig. B.1 Bank system with transactions

2. This could be expressed as

$$accounts{\rightarrow}select(balance < -overdraftLimit)$$

in the *Customer* context.

3. Figure B.1 shows the extended class diagram.

4. This could be:

```
query sumsqdiffs(s1 : Sequence(double),
                 s2 : Sequence(double)) : double
pre: s1.size = s2.size
post:
  result = Integer.subrange(1,s1.size)->collect( i |
                              (s1[i]-s2[i])->sqr())->sum()
```

or equivalently as:

```
static query sumsqdiffs(s1 : Sequence(double),
                        s2 : Sequence(double)) : double
pre: s1.size = s2.size
post:
  result = Integer.sum(1,s1.size,i, (s1[i]-s2[i])->sqr())
```

5. It will converge because each iteration divides the search range in half, so eventually the first condition $ru - rl < tol$ will become true. However it will only find a correct IRR if there is a point r within the initial range $[rl, ru]$ where $price = value(r)$. For example, if the IRR is actually negative (the bond can only make a profit if there is deflation), then searching within the interval [0, 1] will terminate with result 0.

Each iteration of bisection halves the length of the interval being considered. Thus if we start with $rl = 0, ru = 1, r = 0.5$, then after n iterations the interval length will be $\frac{1}{2^n}$.

On termination ru and rl will be within tol of each other (first clause of the function definition). If this occurs after n iterations, the range $ru - rl$ has been divided by 2^n, so that:

$$(ru - rl)/2^n < tol$$
$$(ru - rl)/2^{(n-1)} \geq tol$$

Therefore:

$$-n + log_2(ru - rl) < log_2(tol)$$
$$log_2(ru - rl) + (1 - n) \geq log_2(tol)$$

and:

$$n > log_2(ru - rl) - log_2(tol)$$
$$(n - 1) \leq log_2(ru - rl) - log_2(tol)$$

Therefore we have termination within

$$floor(log_2((ru - rl)/tol)) + 1$$

iterations. For $ru = 1, rl = 0$ this means $floor(-log_2(tol)) + 1$ iterations.

6. Executable specifications can be validated by testing, in addition to validation by inspection/walkthrough.

7. There is less detail to manage at the specification level, so it is simpler to apply the refactorings, in principle. In addition, the refactorings automatically apply to every synthesised implementation via suitable code generators.

Chapter 5

1. This could be:

```
static query coprime(x : int, y : int) : boolean
pre: x > 0 & y > 0
post:
  (x = 1 => result = true) &
  (y = 1 => result = true) &
  (x /= 1 & y /= 1 & x = y => result = false) &
  (x /= 1 & y /= 1 & x /= y => result = (gcd(x,y) = 1))
```

The activity could be:

```
if x = 1 or y = 1 then return true
else if x = y then return false
```

```
else if gcd(x,y) = 1 then return true
else return false
```

2. Applying the formula gives first a 4 year rate of 0.012, then using this with the second bond gives a 5 year rate = 0.0095.

3. The *update* operation would need to adjust the *rate* of each *CashFlow* object *cf* in the investment *flows* sequence, and recalculate *cf.discount* from this:

```
Investment::
update()
post:
  flows->forAll( cf |
             cf.rate = yieldCurve.yield(cf.timePoint) &
             cf.discount = 1/(1 + cf.rate)->pow(cf.timePoint))
```

Here, *cf.rate* is considered to be the effective annual interest rate applying in the period from 0 to *cf.timePoint*. *yieldCurve* is the observed yield curve.

4. The key classes are:

- *Currency*, with a *name* : *String*, such as "Euro", *code* : *String* for a 3-letter code, eg., "EUR", and *symbol* : *String* for the currency symbol, such as "£" for GB pound.
- There are predefined currency subclasses for most of the national currencies, e.g., *USDCurrency* for US dollars, *GBPCurrency* for GB pounds.
- *Money* represents a pair of a currency and a numeric *value* amount.
- *ExchangeRate* associates a numeric *rate* to a pair of currencies, a *source* (conversion from) currency, and a *target* (conversion to) currency. The class provides an *exchange*(*m* : *Money*) : *Money* operation to map an amount *m* in one currency to a corresponding amount in the other. Exchange rates can be chained.
- *ExchangeRateManager*, a Singleton class which stores exchange rates together with a time period for the validity of the rate. The period is defined by a start date and end date.

Figure B.2 shows the structure of this part of QuantLib.

Chapter 6

1. The table would have key based on the *symbol* and *date*. There would be one column family *price* with columns *opening*, *closing*, *high*, *low*, and one column family for *volume* with a single column.

Keys in HBase are in lexicographic order, so the date should be the first part of the key, in order that all records for a single date are stored adjacently. However the given date format will not work correctly for key ordering: for example, "27/12/2017 IBM" will precede "28/12/2017 IBM" but "02/01/2018 IBM" may be listed before both of these. To ensure correct ordering, prepend the date in year:month:day format before the symbol:

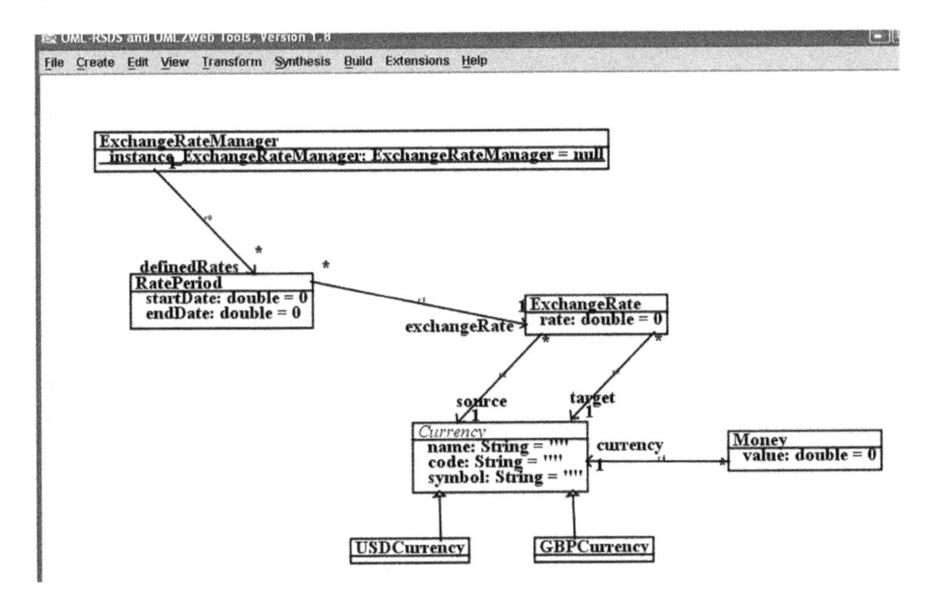

Fig. B.2 FX classes in QuantLib

```
2017:12:29 IBM
2018:01:02 IBM
```

For (i), partition divides the data into chunks for each *map*. The *map* node generates an empty sequence [] for a record that fails the condition $volume > 300000$ and $closing \geq opening + 1$, and generates $[(symbol, date)]$ for a record that satisfies the condition. Shuffle groups all the data with a common symbol into the input for one *reduce*. The *reduce* nodes simply count the size of their input, i.e., the number of different dates that satisfy the condition, and emit that count with the symbol.

For (ii), partition is as for (i). The *map* nodes emit $[(symbol, date, closing - opening)]$ for each record. Shuffle groups these into inputs for each *reduce*, one *reduce* node for each symbol. Each *reduce* node computes the maximum difference value in its inputs, and outputs the sequence of tuples $(symbol, date, diff)$ for which $diff$ is equal to this maximum value.

2. Partition divides the data into chunks for each *map*. The *map* node generates an empty sequence [] for a record that fails the condition $volume > 100000$, and generates $[(symbol, date, closing)]$ for other records. Shuffle groups these into inputs for each *reduce*, one *reduce* node for each symbol. Each *reduce* node computes the maximum closing value in its inputs, and outputs the sequence of tuples $(symbol, date, closing)$ for which $closing$ is equal to this maximum value.

3. For SMA this is:

```
Share::
  adddata(d : ShareDayData) : boolean
```

```
pre: daydata.size > 25 & sma.size > 0
post:
  n = daydata@pre.size &
  prevSMA = sma@pre.last &
  prevD = daydata@pre.last.closing &
  smad = prevSMA + d.closing/26 -
         daydata@pre[n-25].closing/26 &
  daydata->includes(d) &
  sma->includes(smad) &
  (d.closing > smad & prevD <= prevSMA => result = true) &
  (d.closing <= smad or prevD > prevSMA => result = false)
```

For EMA this is:

```
Share::
  adddata(d : ShareDayData) : boolean
  pre: ema.size > 0
  post:
    prevEMA = ema@pre.last &
    prevD = daydata@pre.last.closing &
    emad = prevEMA + alpha*(d.closing - prevEMA) &
    daydata->includes(d) &
    ema->includes(emad) &
    (d.closing > emad & prevD <= prevEMA => result = true) &
    (d.closing <= emad or prevD > prevEMA => result = false)
```

4. 11 of the 17 cases have the required properties, that is 65%.

5. Any associative r can be used, including max, min, sum, prd.

Chapter 7

1. For $t = n * \lambda_1$ we have:

$$y(n * \lambda_1) = \beta_1 + \beta_2 * (1 - exp(-n))/n + \beta_3 * ((1 - exp(-n))/n - exp(-n))$$

At $n = 1$ we have

$$y(\lambda_1) = \beta_1 + \beta_2 * (1 - exp(-1)) + \beta_3 * ((1 - exp(-1)) - exp(-1))$$

So $\beta_1 + 0.632 * \beta_2 + 0.264 * \beta_3 > 0$.

As n becomes large, the coefficients of the β_2 and β_3 factors tend to $\frac{1}{n}$. So $\beta_1 + \beta_2/n + \beta_3/n > 0$ for large n, e.g., $n > 10$.

2. The operation for $slope(n)$ is

```
static query slope(n : int) : double
pre: n > 0 & lambda1 > 0
post:
  result = (yield(n*lambda1) - yield(0))/n
```

3. $yield(0)$ is $\beta_1 + \beta_2$, so

$$\begin{aligned} n * slope(n) = &\beta_2 * (1 - exp(-n) - n)/n + \\ &\beta_3 * ((1 - exp(-n))/n - exp(-n)) \end{aligned}$$

That is:

$$\begin{aligned} slope(n) = &\beta_2 * (1 - exp(-n) - n)/n^2 + \\ &\beta_3 * (1 - exp(-n) - n * exp(-n))/n^2 \end{aligned}$$

Thus for (i) we need $\beta_3 > (n - 1) * \beta_2$, and for (ii) $\beta_3 < (n - 1) * \beta_2$.

Typically β_2 is negative, so (i) will be satisfied for any positive β_3. (ii) requires β_3 to be negative and β_2 positive, or both positive and $\beta_3 < 9 * \beta_2$.

4.

$$N * e^{(n+1)*y(n+1)} = N * e^{n*y(n)} * e^{fr(n)}$$

So $(n + 1) * y(n + 1) = n * y(n) + fr(n)$ and $fr(n) = (n + 1) * y(n + 1) - n * y(n)$.

5.

$$\begin{aligned} fr(t) = &(t + 1) * (\beta_1 + \beta_2 * (1 - exp(-(t + 1)/\lambda_1))/((t + 1)/\lambda_1) + \\ &\beta_3 * ((1 - exp(-(t + 1)/\lambda_1))/((t + 1)/\lambda_1) - exp(-(t + 1)/\lambda_1))) - \\ &t * (\beta_1 + \beta_2 * (1 - exp(-t/\lambda_1))/(t/\lambda_1) + \\ &\beta_3 * ((1 - exp(-t/\lambda_1))/(t/\lambda_1) - exp(-t/\lambda_1))) \end{aligned}$$

The $t * \beta_1$ terms cancel out, as does a term $\beta_2 * \lambda_1$, leaving:

$$\begin{aligned} fr(t) = &\beta_1 + \beta_2 * \lambda_1 * (exp(-t/\lambda_1) - exp(-(t + 1)/\lambda_1)) + \\ &\beta_3 * \lambda_1 * (exp(-t/\lambda_1) - exp(-(t + 1)/\lambda_1)) + \\ &\beta_3 * (t * exp(-t/\lambda_1) - (t + 1) * exp(-(t + 1)/\lambda_1)) \end{aligned}$$

Writing δ for

$$exp(-t/\lambda_1) - exp(-(t + 1)/\lambda_1)$$

gives the simplified equation

$$fr(t) = \beta_1 + \beta_2 * \lambda_1 * \delta + \\ \beta_3 * \lambda_1 * \delta + \beta_3 * (t * \delta - exp(-(t+1)/\lambda_1))$$

6. For the fitting based on yields, the computation of yields from prices is expensive (the IRR computation), however this only needs to be done once, prior to the curve-fitting. In contrast, to perform fitting based on prices it is necessary to compute prices from yields at each fitting attempt, and therefore the computational burden could be higher in this approach.

Chapter 8

1. The exponential terms tend to 1, and I tends to 0, so the value tends to $S - K$.
2.

```
static query euCallOptionPrice(s : double, x : double,
    r : double, q : double,
    dt : double, sigma : double, income : double) : double
pre: sigma > 0 & dt > 0
post:
  p = euPutOptionPrice(s,x,r,q,dt,sigma,income) &
  adjustedX = x*((-r*dt)->exp()) &
  result = p + s - adjustedX
```

3. The options are:

$$b1 : CallOption$$
$$b2 : CallOption$$
$$b1.underlyingAsset = s$$
$$b2.underlyingAsset = s$$
$$b1.amount = b2.amount$$
$$b1.maturity = b2.maturity$$
$$b1.maturityPrice < b2.maturityPrice$$

Where $s : Stock$. The contracts are:

$$con1 : Contract$$
$$con1.holding = b1$$
$$con2 : Contract$$
$$con2.holding = b2$$
$$con1.position = long$$
$$con2.position = short$$

Chapter 9

1. $vdefaults()$ is:

```
CDO::
query vdefaults() : double
pre: n >= 2
post:
  n = loans.size &
  e = edefaults() &
  result = e + n*(n-1)*beta(n) - e*e
```

Where:

```
CDO::
query beta(n : int) : double
pre: n >= 2
post:
  notp = 1 - p &
  notpq = 1 - p*q &
  factor1 = p*p + 2*p*notp*(1 - (1-q)*notpq->pow(n-2)) &
  factor2 = notp*notp*(1 - 2*notpq->pow(n-2) +
            (1 - 2*p*q + p*q*q)->pow(n-2))) &
  result = factor1 + factor2
```

Note the necessary assumption that $n \geq 2$, which was implicit in the corresponding proof in Chap. 9.

2.

```
CDO::
query expectedLoss() : double
post:
  result = sectors->collect( L*sectorLoss())->sum()
```

Where:

```
Sector::
query sectorLoss() : double
post:
  result = Integer.Sum(1,n,m,m*pcond(m)*mu)
```

3. $P(S = 1)$ and $P(S = M)$ are 0, because there is no combination of defaults that can produce a loss of exactly 1 or M: only loss amounts that are multiples of 3 can arise.

4. $pgs(s)$ is $P(S \geq s)$, which is also $1 - P(S < s)$. $P(S < s)$ is the sum of $P(S = l)$ for $l = 0, \ldots, s - 1$, so we can define pgs as:

```
CDO::
query pgs(s : int) : double
post:
  result = 1 - Integer.Sum(0,s-1,l,ps(l))
```

5. The $lossVariance()$ of the portfolio loss is:

```
CDO::
query lossVariance() : double
post:
  result =
     sectors->collect(sectorVarianceContribution())->sum()
```

Where:

```
Sector::
query sectorVarianceContribution() : double
post:
  result = Integer.Sum(1,n,m,m*m*L*L*pcond(m)*mu)
```

The risk contribution RC_k of a sector is then:

```
CDO::
query riskContribution(k : int) : double
pre: 1 <= k & k <= sectors.size
post:
  v = lossVariance() &
  result = sectors[k].sectorVarianceContribution()/v.sqrt
```

6. The simplest approach is to consider that $P(S = s) = 0$ for $s < 0$. This means that ps and vs become:

```
CDO::
query ps(s : double) : double
pre: sectors->forAll( L > 0 )
post:
  (s < 0 => result = 0) &
  (s = 0 =>
     result = (sectors->collect(-mu)->sum())->exp()) &
  (s > 0 =>
     result = (1.0/s)*Integer.Sum(1,sectors.size,k,vs(s,k)))
```

```
CDO::
query vs(s : double, k : int) : double
pre: sectors->forAll( L > 0 )
post:
  Lk = sectors[k].L &
  (s <= 0 => result = 0) &
  (s > 0 =>
     result = Integer.Sum(1,(s/Lk)->floor(),mk,
                          sectors[k].vsk(mk)*ps(s-mk*Lk)))
```

7. This can be defined in terms of the *riskContribution* from Q5:

```
CDO::
query borrowerRiskContribution(k : int, i : int) : double
pre: 1 <= k & k <= sectors.size &
  1 <= i & i <= sectors[k].borrowers.size
post:
  Rk = riskContribution(k) &
  borrowerLoss =
        sectors[k].borrowers[i].omega *
        sectors[k].borrowers[i].L &
  result = (Rk/sectors[k].L)*borrowerLoss
```

8. $s1.L$ is 4.96, $s2.L$ is 3.05, $s1.p$ is 0.0148, $s2.p$ is 0.0165.

Bibliography

1. J. Kirby, *Model-driven Agile Development of Reactive Multi-agent Systems, COMPSAC '06* (2006)
2. K. Lano, The UML-RSDS manual (2018), https://nms.kcl.ac.uk/kevin.lano/umlrsds.pdf
3. J. Letouzey, M. Ilkiewicz, Managing technical debt with the SQALE method. IEEE Softw (2012)
4. Version One, *The 11th Annual State of Agile Report* (2017)

K. Lano and H. Haughton, *Financial Software Engineering*, Undergraduate Topics in Computer Science, https://doi.org/10.1007/978-3-030-14050-2

Index

K. Lano and H. Haughton, *Financial Software Engineering*, Undergraduate Topics in Computer Science, https://doi.org/10.1007/978-3-030-14050-2

Y

Z

CPSIA information can be obtained
at www.ICGtesting.com
Printed in the USA
LVHW100749020919
629413LV00023B/114/P